# EDGE for Mobile Internet

For a listing of recent titles in the *Artech House Mobile Communications Series*, turn to the back of this book.

# EDGE for Mobile Internet

Emmanuel Seurre
Patrick Savelli
Pierre-Jean Pietri

Artech House
Boston • London
www.artechhouse.com

Library of Congress Cataloging-in-Publication Data
A catalog record of this book is available from the Library of Congress.

British Library Cataloguing in Publication Data
Seurre, Emmanuel.
    EDGE for mobile Internet. — (Artech House mobile communications series)
    1. Wireless Internet 2. General Packet Radio Service 3. Global system for mobile
    communications
    I. Title II. Savelli, Patrick III. Pietri, Pierre-Jean
    621.3'845

    ISBN 1-58053-597-6

Cover design by Yekaterina Ratner

Figures 1.26, 1.27, and 3.18: © ETSI 2001. Further use, modification, or redistribution is
strictly prohibited. ETSI standards are available from http://pda.etsi.org/pda/ and http://
www.etsi.org/eds/.

Chapter 6: The OMA logo, Open Mobile Alliance, W@P, W@P Certified, and WAP Forum
marks are worldwide trademarks or registered trademarks of Open Mobile Alliance Ltd.

© 2003 ARTECH HOUSE
685 Canton Street
Norwood, MA  02062

International Standard Book Number:  1-58053-597-6
A Library of Congress Catalog Card Number is available from the Library of Congress.

10 9 8 7 6 5 4 3 2 1

# Contents

# Acknowledgments

The authors would like to express their gratitude to Jacques Achard, David Chappaz, Samuel Rousselin, Solofoniaina Razafindrahaba, Jean-Louis Guillet, and Dominique Cyne for their comments and suggestions concerning the manuscript.

# 1

# GPRS General Overview

The *General Packet Radio Service* (GPRS) allows an end user to send and receive data in packet transfer mode within a *public land mobile network* (PLMN) without using a permanent connection between the *mobile station* (MS) and the external network during data transfer. This way, GPRS optimizes the use of network and *radio resources* (RRs) since, unlike circuit-switched mode, no connection between the MS and the external network is established when there is no data flow in progress. Thus, this RR optimization makes it possible for the operator to offer more attractive fees.

The principles defined for the *Global System for Mobile Communications* (GSM) radio interface were kept for GPRS, since the notions of time slot, frame, multiframe, and hyperframe have not changed for GPRS as compared with GSM. The GPRS standard proposes multislot allocations for data transmission; the network may allocate up to eight time slots per *time division multiple access* (TDMA) frame for a given mobile on uplink and downlink. The GPRS standard proposes four channel coding types allowing throughput per slot ranging from 9.05 Kbps to 21.4 Kbps. This allows a theoretical throughput going up to 171.2 Kbps for data transmission when eight time slots are allocated to the MS.

## 1.1  GPRS Logical Architecture

A strict separation has been defined between the radio and network subsystems. The rationale for this is to reuse the network subsystem with other

1

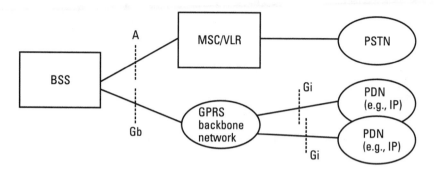

**Figure 1.1**    General architecture of the GPRS network.

radio access technologies such as UMTS. The GPRS network subsystem is also called the GPRS core network or the GPRS backbone network. The GSM network nodes such as *mobile switching center/visitors location register* (MSC/VLR), *home location register* (HLR), and *base station subsystem* (BSS) are reused in the GPRS network architecture. Figure 1.1 gives an overview of the GPRS general architecture.

Two network nodes that are required for packet transfer in the GPRS core network are listed below, as follows:

- *Gateway GPRS support node* (GGSN). The GGSN is a packet router that works with external *packet data networks* (PDNs) and is interfaced with SGSNs via an IP-based GPRS backbone network. A PDN is an external fixed data network such as an Internet network connected to the GPRS network. Packets received from an MS via the SGSN are forwarded by the GGSN to the external PDN as well as the reverse.

- *Serving GPRS support node* (SGSN). The SGSN, which is at the same hierarchical level as the MSC, is the GPRS node serving the MS. It manages GPRS mobility and performs the access control functions so that a user may employ the services provided by a PDN. A SGSN is interfaced with HLR by the *Signaling System No. 7* (SS7) network in order to keep track of the individual MSs' location. It also ensures the routing of packets between the GPRS backbone network and the radio subsystem. Figure 1.2 shows the general architecture of the GPRS backbone network.

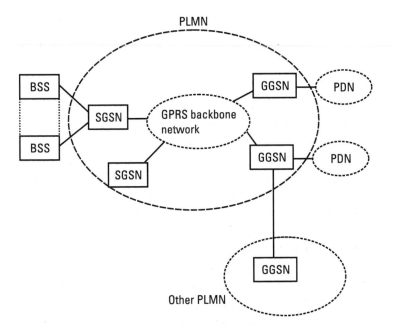

**Figure 1.2** General architecture of the GPRS backbone network.

Other equipment used on existing GSM network has evolved to support data transmission in packet-switched mode. This equipment is listed as follows:

- *BSS.* The BSS is enhanced in order to handle the GPRS functions over the radio interface (e.g., new packet channels). A *packet control unit* (PCU) has been defined in the BSS to serve the GPRS functions.

- *MSC/VLR.* The MSC/VLR can be enhanced to coordinate the GPRS and non-GPRS services during the paging procedure for circuit-switched calls and during GPRS and non-GPRS location update procedures. This coordination takes place only when the Gs interface between the MSC/VLR and SGSN is present.

- *HLR.* The HLR has been updated in order to handle GPRS subscriber information and GPRS MS location information.

New interfaces are defined between the different network elements. These interfaces are standardized to allow interoperability between network nodes that are provided by different manufacturers in one network. These interfaces are as follows:

- *Gb interface.* The Gb interface is located between the SGSN and the BSS. It supports both signaling and data transfer. It is used for packet transfer, cell reselection, and such.

- *Gn/Gp interface.* The Gn/Gp interface is defined between GPRS support nodes in the GPRS core network. It is used for the transfer of packets and signaling between the GSNs. The Gn interface is defined between two GSNs (SGSN or GGSN) within the same PLMN, whereas the Gp interface is defined between two GSNs located in different PLMNs.

- *Gs interface.* The Gs interface is located between the MSC/VLR and the SGSN. Through this interface an association is created between the SGSN and the MSC/VLR to coordinate MSs that are both GPRS-attached and IMSI-attached for circuit-switched paging and for combined location procedures.

- *Gr interface.* The Gr interface between the SGSN and the HLR is used to retrieve or update the GPRS subscriber profile and location during *GPRS mobility management* (GMM) procedures.

- *Gf interface.* The Gf interface between the SGSN and the *equipment identity register* (EIR) allows verification of the terminal's identity.

- *Gc interface.* The Gc interface is defined between the GGSN and the HLR. It is used to retrieve routing information needed to forward incoming packets from the PDN to the SGSN serving the mobile for which it is intended.

- *Gi interface.* The Gi interface is located between the GGSN and the external PDN. The protocols that are involved in this interface are dependent on the external PDN. The *Internet Protocol* (IP) is supported by this interface, but the *Point-to-Point* (PTP) Protocol may also be supported.

Figure 1.3 shows the different elements of a GPRS network together with their associated interfaces.

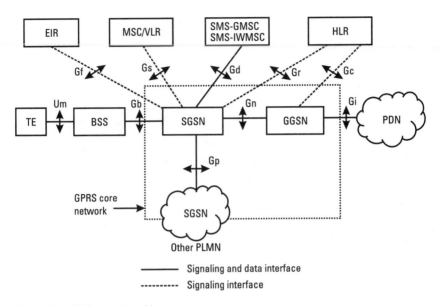

**Figure 1.3** GPRS network architecture.

## 1.2 Transmission and Signaling Planes

A complex and distributed network architecture such as GPRS is made up of a transmission plane and a signaling plane. The transmission plane or user plane provides the means of transmission for user information transfer between the MS and an external packet-switched network. The signaling plane controls and supports the transmission plane functions within the network.

### 1.2.1 Transmission Plane

The transmission plane consists of a layered protocol structure providing user data transfer. Despite the various interfaces across the GPRS network, an end-to-end transmission path is to be ensured according to information transfer control procedures (e.g., flow control, error detection, error correction, and error recovery). The transmission plane in the network subsystem is independent of the one defined in the radio subsystem according to the Gb interface. Figure 1.4 shows the layered protocol structure in the transmission plane between the MS and the GGSN.

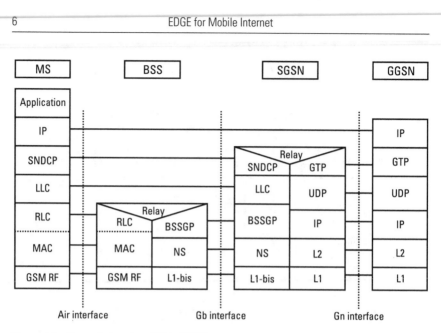

**Figure 1.4**  Transmission plane MS to GGSN.

The GSM *radio frequency* (RF) layer is split into two sublayers—physical RF layer and physical link layer. The physical RF layer is used to control physical channels, (de)modulation, transmission, and reception of blocks on the radio interface. The physical link layer is used to control channel coding, interleaving, power control, measurements, and synchronization.

The *medium access control* (MAC) layer is used to control access to the radio channel between the mobiles and the network.

The *radio link control* (RLC) layer adapts the *protocol data unit*(s) (PDU) received from the *logical link control* (LLC) layer to the RLC data transport unit. The RLC segments the LLC PDUs into RLC data blocks and reassembles them in the reverse direction. It provides retransmission mechanisms for erroneous data blocks.

The LLC layer provides a reliable ciphered link between the MS and the SGSN. This link is independent of the underlying layers.

The purpose of the *Subnetwork Dependent Convergence Protocol* (SNDCP) layer is to map the IP layer with the underlying transport network. Compression, segmentation and, multiplexing of network layer messages are also performed by the SNDCP layer.

The *Base Station Subsystem GPRS Protocol* (BSSGP) in the transmission plane controls the transfer of LLC frames across the Gb interface.

The *network service* (NS) layer is based on *frame relay* (FR) between the BSS and SGSN. It conveys BSSGP PDUs.

The *GPRS Tunneling Protocol* (GTP) for the user plane (GTP-U) provides services for carrying a user data packet between the GPRS support nodes within the GPRS backbone network.

The *User Datagram Protocol* (UDP) conveys GTP PDUs in the GPRS backbone network.

The IP is used to route user data within the GPRS backbone network.

Two relays functions are implemented in the transmission plane. The relay function in the BSS forwards the LLC PDUs between the air interface and the Gb interface, while the relay function in the SGSN forwards the *Packet Data Protocol* (PDP) PDUs between the Gb and Gn interfaces.

## 1.2.2 Signaling Plane

The signaling plane enables performance of the following functions:

- *GPRS network access connection.* This is a function that provides the user with a means to use GPRS services. A set of procedures is defined to control the access connection (e.g., IMSI attach for GPRS services, IMSI detach for GPRS services).

- *External network access connection.* This is a function that allows control of the attributes of an established network access connection by activating, deactivating, or modifying a context between the MS, the SGSN, and the GGSN.

- *Mobility management.* This is a function that ensures the continuity of packet services within the PLMN or within another PLMN by keeping track of the current MS location.

- *Adaptation of network resources.* This is a function that calculates the amount of network resources required for the requested *quality of service* (QoS).

Figure 1.5 shows the signaling plane between the MS and the SGSN.

The GMM layer manages the procedures related to GPRS mobility between the MS and SGSN.

The *session management* (SM) layer manages the procedures related to the contexts between the MS, the SGSN, and the GGSN.

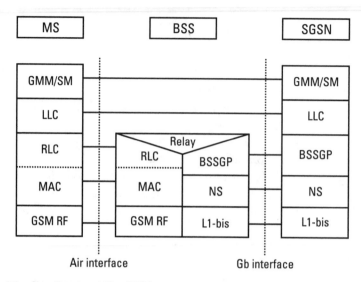

**Figure 1.5**    Signaling plane MS to SGSN.

The BSSGP in the signaling plane provides functions associated with mobility management between an SGSN and a BSS. Figure 1.6 shows the signaling plane between two GSNs.

The *GTP for the control plane* (GTP-C) tunnels signaling messages between GPRS support nodes in the GPRS backbone network. The *GPRS support nodes* (GSNs) of the GPRS backbone network are interfaced with SS7 network in order to exchange information with GSM SS7 network nodes such as HLR, MSC/VLR, EIR, and SMS-GMSC. These new interfaces are listed in Table 1.1.

**Figure 1.6**    Signaling plane GSN to GSN.

**Table 1.1**
New Interfaces with the SS7 Network

| Interface Name | Location | Mandatory or Optional |
|---|---|---|
| Gr | SGSN—HLR | Mandatory |
| Gc | GGSN—HLR | Optional |
| Gf | SGSN—EIR | Optional |
| Gd | SGSN—SMS GMSC or SGSN—SMS IWMSC | Optional |
| Gs | SGSN—MSC/VLR | Optional |

## 1.3 The Radio Interface

### 1.3.1 Physical Layer

The GPRS physical layer relies on the same underlying principles as GSM. It is based on a combination of TDMA and *frequency division multiple access* (FDMA). Frequency channels are 200 kHz wide; the TDMA frame lasts 4.615 ms and consists of eight time slots. As for GSM, the physical channels are defined by a frequency channel and time slot pairing for the uplink and downlink paths (see Figure 1.7), and logical channels are mapped onto the

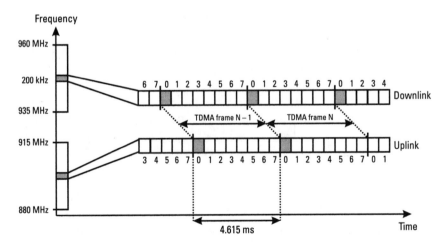

**Figure 1.7**  Combination of FDMA and TDMA.

physical channels for data traffic and for signaling. As shown in this section, new logical channels have been defined for GPRS.

Further, many characteristics differ from the GSM circuit-switched services, such as the use of the 52-multiframe (instead of the 26-multiframe in GSM traffic), new *coding schemes* (CSs), and new power control algorithms for uplink and downlink. Moreover, a link adaptation mechanism is used to change the CS according to the radio conditions in order to find the best trade-off between error protection and achieved throughput.

At the RF physical layer, the main characteristic is the possibility to allocate several physical channels to a given MS to provide higher data rate packet services. This means that an MS can receive or transmit data on several time slots per TDMA frame.

### 1.3.1.1 Definition of the Physical Channel

The GPRS multiframe length is 52 TDMA frames; it contains 12 blocks (B0 to B11) of 4 consecutive TDMA frames plus 4 idle frames (see Figure 1.8). A physical channel is referred to as a *packet data channel* (PDCH). It may be fully defined by a frequency and time slot pairing (one time slot in downlink and the corresponding time slot in uplink). On a given PDCH, blocks of 4 bursts, called radio blocks, are used to convey the logical channels, transmitting either data or signaling.

### 1.3.1.2 Packet Data Logical Channels

Packet data logical channels, defined for GPRS data traffic and signaling, are mapped on top of physical channels. There are two types of logical channels: traffic channels and control channels. Among the control channels, three

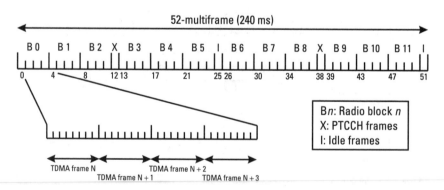

**Figure 1.8**   The 52-multiframe.

subtypes have been defined for GPRS—broadcast, common control, and associated. In addition to the GPRS logical channels, the GSM control channels (BCCH, CCCH, RACH) are used for the MS access to the network and for the packet transfer establishment when GPRS control channels are not allocated in a GPRS cell.

The different packet data logical channels are as follows:

- *Packet data traffic channel* (PDTCH). The PDTCH is the channel on which the user data is transmitted during uplink or downlink packet transfer. It is a unidirectional channel, either uplink (PDTCH/U) for a mobile originated packet transfer or downlink (PDTCH/D) for a mobile terminated packet transfer.

- *Packet associated control channel* (PACCH). The PACCH is a unidirectional channel that is used to carry signaling for a given MS during uplink or downlink packet data transfer. It is always associated with one or several PDTCHs allocated to an MS.

- *Packet broadcast control channel* (PBCCH). The PBCCH broadcasts information on the cell the MS is camping on (the cell that is selected by the MS) and on neighbor cells. It contains the parameters needed by the mobile to access the network. When there is no PBCCH in the cell, the information is broadcast on BCCH.

- *Packet common control channel* (PCCCH). The PCCCH is a set of logical channels composed of PRACH, PPCH, and PAGCH:

  - *Packet random access channel* (PRACH) is used by the MS to initiate an uplink access to the network.

  - *Packet paging channel* (PPCH) is used by the network to page the MS in order to establish a downlink packet transfer.

  - *Packet access grant channel* (PAGCH) is used by the network to assign RRs to the mobile for a packet transfer.

  PCCCH is present in the cell only if PBCCH is present. If it is not present, the common control signaling for GPRS is handled through the GSM *common control channels* (CCCHs).

- *Packet timing advance control channel* (PTCCH). The PTCCH is a bidirectional channel that is used to adaptively update the MS time synchronization information [*timing advance* (TA)]. It is mapped on frame numbers 12 and 38 of the 52-multiframe, as shown in Figure 1.8.

**Table 1.2**
Summary of the Various GPRS Logical Channels

| Logical Channel | Abbreviation | Uplink/ Downlink | Task |
|---|---|---|---|
| Packet broadcast control channel | PBCCH | DL | Packet system information broadcast |
| Packet paging channel | PPCH | DL | MS paging for downlink transfer establishment |
| Packet random access channel | PRACH | UL | MS random access for uplink transfer establishment |
| Packet access grant channel | PAGCH | DL | Radio resource assignment |
| Packet timing advance control channel | PTCCH | UL/DL | Timing advance update |
| Packet associated control channel | PACCH | UL/DL | Signaling associated with data transfer |
| Packet data traffic channel | PDTCH | UL/DL | Data channel |

Table 1.2 provides a summary of the GPRS logical channels.

We will not discuss here how the different logical channels are mapped onto the 52-multiframe physical channels. It is nevertheless important to note that this mapping can be dynamically configured by the network. This allows the system to adapt to the network load by allocating or releasing resources whenever needed. Further information regarding this topic may be found in [1].

### 1.3.1.3   Definition of the Multislot Classes

For the higher data rates, a GPRS MS may support the use of multiple PDCHs per TDMA frame. The maximum number of time slot that may be allocated to the mobile on the uplink and on the downlink depends on the MS multislot capability. Multislot classes are defined specifying for a mobile the maximum number of time slots in *reception* (Rx) and the maximum number of time slots in *transmission* (Tx). Thus, the number of used time slots may be different in uplink and in downlink, for asymmetrical services.

In addition, a limit is specified in each multislot class for the total of received and transmitted time slots (Sum) supported by the MS per TDMA frame. The multislot class of the MS is sent to the network during the GPRS attach procedure. Table 1.3 lists the MS multislot classes. Type 1 MSs cannot transmit and receive at the same time, but type 2 MSs can.

**Table 1.3**
Mobile Multislot Classes

| Multislot Class | Maximum Number of Time Slots | | | Type |
| | Rx | Tx | Sum | |
| --- | --- | --- | --- | --- |
| 1 | 1 | 1 | 2 | 1 |
| 2 | 2 | 1 | 3 | 1 |
| 3 | 2 | 2 | 3 | 1 |
| 4 | 3 | 1 | 4 | 1 |
| 5 | 2 | 2 | 4 | 1 |
| 6 | 3 | 2 | 4 | 1 |
| 7 | 3 | 3 | 4 | 1 |
| 8 | 4 | 1 | 5 | 1 |
| 9 | 3 | 2 | 5 | 1 |
| 10 | 4 | 2 | 5 | 1 |
| 11 | 4 | 3 | 5 | 1 |
| 12 | 4 | 4 | 5 | 1 |
| 13 | 3 | 3 | N/A | 2 |
| 14 | 4 | 4 | N/A | 2 |
| 15 | 5 | 5 | N/A | 2 |
| 16 | 6 | 6 | N/A | 2 |
| 17 | 7 | 7 | N/A | 2 |
| 18 | 8 | 8 | N/A | 2 |
| 19 | 6 | 2 | N/A | 1 |
| 20 | 6 | 3 | N/A | 1 |
| 21 | 6 | 4 | N/A | 1 |
| 22 | 6 | 4 | N/A | 1 |
| 23 | 6 | 6 | N/A | 1 |

N/A: Not applicable

**Table 1.3** (continued)

| Multislot Class | Maximum Number of Time Slots | | | Type |
| | Rx | Tx | Sum | |
| --- | --- | --- | --- | --- |
| 24 | 8 | 2 | N/A | 1 |
| 25 | 8 | 3 | N/A | 1 |
| 26 | 8 | 4 | N/A | 1 |
| 27 | 8 | 4 | N/A | 1 |
| 28 | 8 | 6 | N/A | 1 |
| 29 | 8 | 8 | N/A | 1 |

N/A: Not applicable

### 1.3.1.4 Channel Coding

Four CSs, CS-1 to CS-4, have been defined for GPRS, offering a decreasing level of protection. The coding rate is the lowest with CS-1 (maximum redundancy) and is the highest for CS-4 (no redundancy). The CS to be used is chosen by the network according to the radio environment. This mechanism is called link adaptation (see Section 1.3.1.5). The coding is based on a *cyclic redundancy code* (CRC), followed by a convolutional encoding, for CS-1 to CS-3. There is only a CRC for CS-4. Puncturing is applied to adapt the convolutional encoder output to the radio block length. Finally, block interleaving over the radio block makes it possible to improve the decoding performance at the receiver. The principle for the coding of one radio block for CS-1 to CS-3 is shown in Figure 1.9.

The mobile always transmits with a CS ordered by the network, whereas in Rx the mobile performs a blind detection of the used CS. This detection is done by analyzing the stealing flags (8 bits per radio block, at the extremities of the training sequences), one different stealing flag pattern being defined for each of the CSs.

A summary of the four CS characteristics is given in Table 1.4. This table specifies the total coding rate for each CS and for a radio block; they are as follows:

- The pre-encoding of the *uplink state flag* (USF) field;
- The length of the data to be encoded;

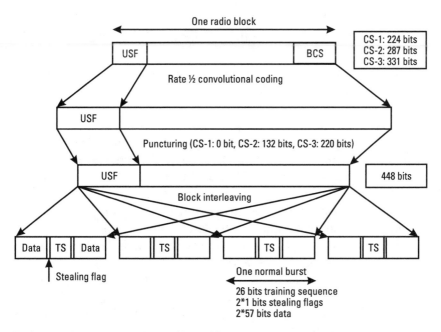

**Figure 1.9** Radio block encoding for CS-1 to CS-3.

- The *block check sequence* (BCS), which is the CRC field;
- The number of tail bits (used to improve the decoding performance);
- The number of bits after the encoding;
- The number of punctured bits.

**Table 1.4**
Coding Parameters for the GPRS Coding Schemes

| Scheme | Code Rate | USF | Pre-coded USF | Radio Block excl. USF and BCS | BCS | Tail Bits | Coded Bits | Punctured Bits | Data Rate (Kbps) |
|---|---|---|---|---|---|---|---|---|---|
| CS-1 | 1/2 | 3 | 3 | 181 | 40 | 4 | 456 | 0 | 9.05 |
| CS-2 | ≈2/3 | 3 | 6 | 268 | 16 | 4 | 588 | 132 | 13.4 |
| CS-3 | ≈3/4 | 3 | 6 | 312 | 16 | 4 | 676 | 220 | 15.6 |
| CS-4 | 1 | 3 | 12 | 428 | 16 | — | 456 | — | 21.4 |

*From:* [2].

Finally, the data rate is given in the last column. This figure corresponds to the ratio of data bits at the encoder input to the duration of a radio block (20 ms).

Note that GPRS signaling is always sent with the CS-1, the other CSs being used only on the PDTCH.

### 1.3.1.5   Link Adaptation

The basic principle of link adaptation consists of changing the CS used for transmission according to the radio conditions. When the radio conditions are bad, the level of protection is increased by the use of a lower code rate. Similarly, when the radio conditions are good, the level of protection is decreased. This allows the best trade-off between error protection and achievable data rate. If, for instance, the C/I is high, a low level of protection is applied to achieve a high data rate. If the C/I is low, the level of protection is increased, which leads to a lower data rate.

The criteria for good or bad radio conditions (Doppler shift due to the mobile speed, multipath, interference, and so on) are determined by the network, based on the measurements that are performed by the MS in downlink or by the BTS in uplink.

### 1.3.1.6   Principles of Power Control

Power control is often used in wireless communications systems to reduce the interference to the other users (which improves the spectrum efficiency), while keeping a good quality of radio link and reducing the power consumption in the MS. It consists of adapting the level of transmitted signal in uplink and in downlink to the propagation conditions. In GSM, the power level used by the MS transmitter is commanded by the network, based on measurements of BTS receive signal strength in uplink. In GPRS, since there is not necessarily a continuous two-way connection, uplink power control is either performed by the mobile itself, based on *received signal level* (RXLEV) measurements, or performed indirectly, based on BSS power control orders. The first type of power control is called open-loop power control, and the second one closed-loop power control. Also, a combination of open-loop and closed-loop power control may be used. In the downlink, power control is performed by the BTS, based on measurement reports sent by individual mobiles.

1.3.1.7   Radio Environment Monitoring

Several types of radio measurements are performed by the MS, which uses them to compute its transmission power (in open-loop power control), and for cell selection and reselection.

Measurements are also reported to the network, for RLC purposes. The various measurements are as follows:

- *RXLEV.* These measurements of the RXLEV are performed on the BCCH channel of the serving cell and of the neighbor cells for the purpose of cell reselection. In packet transfer mode, the serving cell RXLEV measurement can also be used for downlink CS adaptation, network controlled cell reselection (see Section 1.3.2.6), and power control in uplink and downlink. These measurements are made on the BCCH because it is transmitted at a constant output power, at the maximum BTS level. It is therefore suitable for an accurate estimation of downlink path loss.

- *Quality measurements* (RXQUAL). The RXQUAL measurements consist of estimations of the average *bit error rate* (BER) before channel decoding. They are computed only in packet transfer mode, on the downlink blocks that the mobile receives. The network uses the RXQUAL reports for network controlled cell reselection, dynamic CS adaptation, and downlink power control. The estimation is obtained by averaging the BER on the successfully decoded blocks (that is, on blocks where no error is detected by the CRC check) intended for the MS.

- *Interference measurements.* These measurements correspond to a RXLEV estimation, performed on a frequency that is different from the BCCH of the serving cell. The goal of these measurements is for the network to have an estimation of the interference level due to other cells on a given PDTCH. This information may, for instance, be used to optimize mobile RR allocation, CS adaptation, power control, or network-controlled cell reselection, or simply to collect network statistics.

Note that RXLEV and RXQUAL measurements may also be performed at the BSS side for each MS. They are used for network-controlled cell reselection, uplink power control, and dynamic CS adaptation.

## 1.3.2   Radio Resource Management (RRM)

This section describes the RLC/MAC layer. It gives the main principles of this layer and the way radio resources are allocated to the mobile and data are exchanged between the network and the mobile.

### 1.3.2.1   Basic Principles of RRM

This section details various fundamental concepts that are used for RRM. Whether the mobile is transmitting (or receiving) packets or not, it performs different actions that are based on two RR states. These two RR operational states are described next.

During packet transfer, different mobiles can be multiplexed by the network on the same physical channel. The downlink multiplexing is performed directly by the network that addresses radio blocks to the selected mobile. All the mobiles that are sharing the same downlink PDCH, decode all the radio blocks. An identifier that is assigned during resource allocation is used to discriminate the radio blocks addressed to a given mobile. On the uplink side, the multiplexing is also controlled by the network, but in this case the network has to assign uplink radio block occurrences. The second section describes the mechanisms that can be used by the network to perform this multiplexing.

The third section gives a description of the broadcast channels that are used by the network to broadcast information related to the cell to the different mobiles that are located within it.

The last section describes the RLC/MAC block format that is used as the basic transport unit on the radio interface.

### RR Operating Modes

Two operating states have been defined at the RR level: packet idle mode and packet transfer mode. Each of these states characterizes the RR activity of the MS.

In packet idle mode, the mobile has no RR allocated. The mobile leaves this state when upper layers request the transfer of uplink data. In this case, the mobile enters a transitory state before going into packet transfer mode. The switch to packet transfer mode occurs at the end of the contention resolution phase when the mobile has been uniquely identified at network side. The mobile also leaves the packet idle mode when it receives from the network a downlink resource allocation command. In this case, the mobile enters directly the packet transfer mode state. In packet idle mode, the MS performs paging and broadcast information listening.

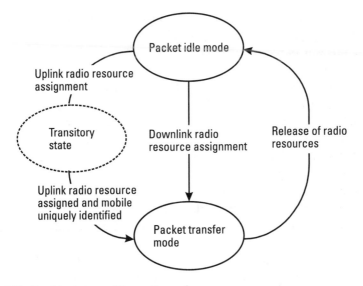

**Figure 1.10**  Transition between RR operating modes.

In packet transfer mode, the mobile has been allocated either uplink RRs or downlink RRs or both by the network. Departure from the packet transfer mode state occurs when the RRs are released. This is the case at the end of a packet transfer, at a radio link fails failure, or when the mobile initiates a cell reselection toward a new cell. Figure 1.10 shows the RR transition state diagram.

*Allocation Modes on the Uplink*

In order to share the uplink bandwidth between the different mobiles mapped on the same PDCH, and to allocate an uplink radio block instance to a particular mobile, different allocation schemes have been defined (e.g., dynamic allocation, extended dynamic allocation, and fixed allocation). The dynamic allocation and the fixed allocation are mandatory on the mobile side whereas extended dynamic allocation is optional. On the network side, there is no particular requirement.

The principle of dynamic allocation is to allow uplink transmission to mobiles sharing the same PDCH dynamically, on a block-by-block basis. On the other hand, the principle of fixed allocation is very simple. It consists of indicating to the mobile during the resource allocation or reallocation phase or fixed block occurrences on the allocated PDCHs in which the mobile is allowed to transmit.

In the following section the principle of dynamic allocation is described in greater detail because of its higher complexity and its exclusive employment by most BSS manufacturers.

During the resource assignment for an uplink transfer, a USF is given to the MS for each allocated uplink PDCH. This USF is used as a token given by the network to allow transmission of one uplink radio block.

In order to allocate one radio block occurrence on one uplink PDCH, the network includes, on the associated downlink PDCH, the USF in the radio block immediately preceding the allocated uplink block occurrence. When the mobile decodes its assigned USF value in a radio block sent on a downlink PDCH, it transmits an uplink radio block in the next uplink radio block occurrence—that is, $B(x)$ radio block if the USF was detected in $B(x-1)$ radio block. The principle of dynamic allocation is illustrated in Figure 1.11.

The USF is included in the header of each downlink RLC/MAC block. Dynamic allocation requires the decoding of all the downlink blocks sent on the allocated PDCHs. The USF coding (3 bits) enables the multiplexing of eight mobiles on the same uplink PDCH.

Dynamic allocation can also be used in such a way that the decoding of one USF value allows the sending of four consecutive uplink blocks on the same PDCH. A concept of USF granularity is used to indicate the number of uplink radio blocks (one or four) that can be sent upon detection of an assigned USF value. The USF granularity is signaled during the uplink RR allocation by the network.

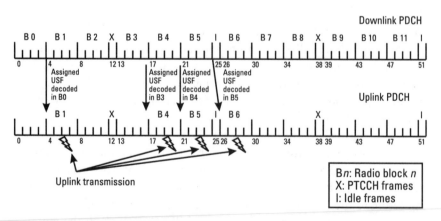

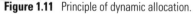

**Figure 1.11**  Principle of dynamic allocation.

*Broadcast Information Management*

In each cell, two channels are dedicated to the broadcast of information relative to the serving cell and the neighbor cells. The first one is the *broadcast control channel* (BCCH), and the second the PBCCH. Note that the PBCCH is optional. It is used to broadcast GPRS information. The BCCH is multiplexed on the time slot 0 of the carrier that transmits the FCH and the SCH. The location of the PBCCH is indicated within one message that is broadcast on BCCH.

The parameters broadcast on these channels are the list of frequencies that are used in the cell, the neighbor cell frequencies, the GSM and GPRS logical channel description, and the access control parameters. The mobile uses the broadcast serving cell frequencies to derive its frequency allocation during resource assignment. It uses the neighbor cell frequencies for measurement and cell reselection purposes. The logical channel description indicates how the different logical channels on the time slots are multiplexed. The network broadcasts access control parameters and puts constraints on the access channels in order to avoid congestion.

The serving cell and neighbor cell parameters are broadcast within messages called *SYSTEM INFORMATION* (SI) messages on BCCH and *PACKET SYSTEM INFORMATION* (PSI) messages on PBCCH. Based on this information the MS is able to decide whether and how it may gain access to the system via the cell on which it is camping.

The SI and PSI messages are cyclically broadcast within the cell. Each MS has to periodically decode the SI and PSI messages in order to detect any change in the cell configuration.

*RLC/MAC Block Formats*

As seen previously in this chapter, the radio block is an information block transmitted over four consecutive bursts on a given PDCH. The RLC/MAC block is transmitted in a radio block to carry data and RLC/MAC signaling.

RLC data blocks are transmitted on the PDTCH, and RLC/MAC control blocks are transmitted on the signaling channels PACCH, PCCCH, and PBCCH.

A MAC header is present in each type of radio block, described as follows:

- *The control block.* The RLC/MAC block consists of a MAC header and an RLC/MAC control block as shown in Figure 1.12. This

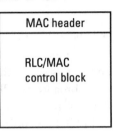

**Figure 1.12** RLC/MAC block structure for control messages.

block is always encoded using CS-1. The size of the RLC/MAC control block is 22 bytes and the size of the MAC header is 1 byte.

• *The RLC data block.* The RLC data block consists of an RLC header, an RLC data unit, and spare bits as shown in Figure 1.13.

Depending on the channel coding (CS-1, CS-2, CS-3, CS-4), a block contains 184, 271, 315, or 431 bits, including the MAC header. The number of spare bits is 0, 7, 3, 7 for, respectively, CS1, CS2, CS3, and CS4.

### 1.3.2.2 Packet Transfer Management

This section details how the transfer of packets is managed at the RLC/MAC layer. The first section deals with the terminology that is used to name and identify a packet transfer. The second section details some procedures that are used to allocate uplink or downlink resources to the MS. The third section deals with the RLC principles that are used to transfer data packet. The fourth section explains the release of the RRs.

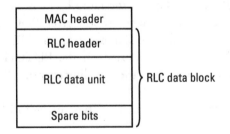

**Figure 1.13** RLC/MAC block structure for data transfer.

## Temporary Block Flow (TBF) Definition

A TBF is a physical connection that is established between the mobile and the BSS at the RR level. This connection is used to transfer packets over the radio interface in one direction. When two transfers in opposite directions occur for the same MS, one uplink TBF and one downlink TBF are established at the same time. The TBF is established for the duration of the transfer. Once no more LLC frames need to be transferred, it is released.

Note that there is at most one TBF established per MS and per direction. When there is at the same time an uplink TBF and a downlink TBF established for the same MS, the TBFs are called concurrent.

A TBF can be mapped over several PDCHs; TBFs belonging to different MSs can share the same or a group of common PDCHs (GPRS multiplexing principle).

Each TBF is identified by a *temporary flow identifier* (TFI) assigned by the network. So in case of concurrent TBFs, one TFI identifies the uplink TBF and another one the downlink TBF. The TFI is used to differentiate TBFs sharing the same PDCHs in one direction.

## RR Allocation

This section briefly describes the different GPRS resource allocation procedures that can use the network to establish a TBF. Two kinds of procedures are used—the procedures for uplink TBF establishment and for downlink TBF establishment. The network allocates the resources using the PCCCH, if present in the cell; otherwise, this is done using the CCCH.

*Uplink TBF Establishment* The mobile triggers the establishment of an uplink TBF for the following reasons:

- To perform an uplink data transfer;
- To answer to a paging;
- To perform a GMM procedure (e.g., routing area update procedure, GPRS attach procedure) or SM procedure (e.g., PDP context activation procedure).

Two different procedures have been defined for the establishment of an uplink TBF. The one-phase access procedure is the basic and fastest way to request an uplink TBF in RLC acknowledged mode in one phase. The mobile requests a two-phase access procedure in order to establish a TBF in two phases. It is also possible to establish an uplink TBF during a downlink TBF.

The mobile requests the establishment of an uplink TBF by sending a CHANNEL REQUEST message on RACH or a PACKET CHANNEL REQUEST message on PRACH. These messages are sent within one access burst. Two different formats of access burst on PRACH are defined. The first one contains 8 bits information and uses the same coding as on RACH. The second one contains 11 bits of information allowing the transmission of more details on the requested TBF.

*One-Phase Access Procedure*   Figure 1.14 illustrates the scenario for a TBF establishment in one-phase access on CCCH. This procedure is used when there is no PCCCH in the cell.

The mobile triggers the procedure by sending a CHANNEL REQUEST message on the RACH indicating one-phase access request. Upon receipt of this message the network allocates one uplink PDCH to the mobile and direct uplink resources if the network has implemented fixed allocation or an USF value if dynamic allocation is used.

The network can allocate only one PDCH to the mobile because of the impossibility of signaling the multislot class within the CHANNEL REQUEST message and because of a limitation in the length of the IMMEDIATE ASSIGNMENT message.

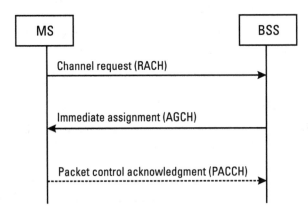

**Figure 1.14**   One-phase access establishment scenario on CCCH.

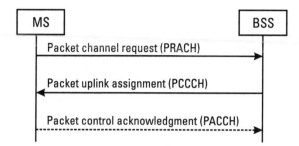

**Figure 1.15** One-phase access establishment scenario on PCCCH.

The network may request an acknowledgment from the mobile. The acknowledgment is requested by setting a polling bit in the IMMEDIATE ASSIGNMENT message. In this case the mobile sends a PACKET CONTROL ACKNOWLEDGMENT message on the assigned PDCH.

Figure 1.15 illustrates the scenario for one-phase access uplink establishment on PCCCH.

This procedure is almost the same as the one on CCCH. The mobile requests the establishment of an uplink TBF by sending a PACKET CHANNEL REQUEST on PRACH. This message contains the multislot class of the mobile. The network is then able to provide RRs on multiple PDCHs within the PACKET UPLINK ASSIGNMENT message. If the network requests the transmission of an acknowledgment, the mobile will send a PACKET CONTROL ACKNOWLEDGEMENT message on PACCH.

*Two-Phase Access Procedure*    Figure 1.16 describes the procedure used to establish an uplink TBF in two-phase access on CCCH.

The mobile initiates the two-phase access procedure by sending a CHANNEL REQUEST message on RACH requesting a two-phase packet access. Upon receipt of the CHANNEL REQUEST message, the network sends an IMMEDIATE ASSIGNMENT message to the mobile on AGCH. This message allocates a single uplink block occurrence in which the mobile sends the PACKET RESOURCE REQUEST message.

Within this message the mobile is able to specify its complete radio access capabilities (multislot class, maximum output power, frequency band supported) and QoS parameters relative to the LLC frame to transmit. The network can allocate uplink resources to the mobile by taking into account all these information by sending a PACKET UPLINK ASSIGNMENT

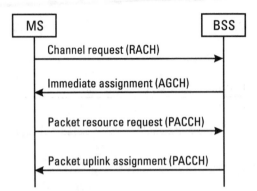

**Figure 1.16**  Two-phase access establishment scenario on CCCH.

message on PACCH. This message contains all the parameters necessary to transfer uplink RLC blocks.

Figure 1.17 illustrates the scenario for two-phase access uplink TBF establishment on PCCCH when the mobile is in packet idle mode.

This procedure is the same as the previous one except that a PACKET CHANNEL REQUEST message is sent on PRACH instead of the CHANNEL REQUEST message and the uplink block occurrence is assigned by sending the PACKET UPLINK ASSIGNMENT message on PCCCH.

*Assignment of Uplink Resources During a Downlink TBF*  The mobile can to establish an uplink TBF when it is in packet transfer mode—thus, during a downlink TBF. Figure 1.18 shows this procedure.

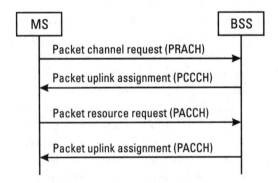

**Figure 1.17**  Two-phase access establishment scenario on PCCCH.

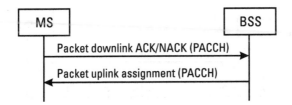

**Figure 1.18**  Procedure for uplink establishment when the MS is in packet transfer mode.

The mobile requests the establishment of an uplink transfer during a downlink TBF with the PACKET DOWNLINK ACK/NACK message. This message is used to acknowledge the RLC data blocks received during the downlink transfer.

*Downlink TBF Establishment*  Figure 1.19 shows an example of downlink TBF establishment on CCCH when the mobile is in packet idle mode. When the network initiates this procedure, the location of the MS is known at cell level. The BSS is able to directly allocate downlink RRs.

When the network receives a downlink LLC PDU to transmit to the mobile, it initiates the establishment of a downlink TBF by sending an IMMEDIATE ASSIGNMENT message to the MS on CCCH. This message is sent on any block of the CCCH if the mobile is in non-DRX mode; otherwise it is sent on one block corresponding to the paging group of the mobile.

Because of a limitation of the IMMEDIATE ASSIGNMENT message length, the BSS is not able to allocate more than one time slot in downlink to the MS despite knowing its multislot class. However, once the resource has been allocated, the network will be able to send a PACKET DOWN-LINK ASSIGNMENT message on PACCH that could allocate several downlink PDCHs to the mobile.

Figure 1.20 shows an example of downlink TBF establishment on PCCCH when the mobile is in packet idle mode.

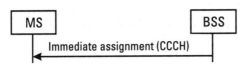

**Figure 1.19**  Downlink TBF establishment on CCCH.

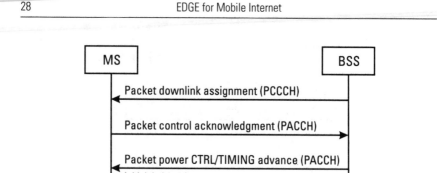

Figure 1.20  Example of downlink TBF establishment on PCCCH.

When the network receives a downlink LLC PDU to transmit to the mobile, it initiates the establishment of a downlink TBF by sending a PACKET DOWNLINK ASSIGNMENT message to the MS on PCCCH.

This message is sent on any block of the PCCCH where paging may appear if the mobile is in non DRX mode; otherwise, it is sent on one block corresponding to the paging group of the mobile.

In order to provide a TA value allowing the MS to transmit, the network may request the sending of an acknowledgment message after receiving the PACKET DOWNLINK ASSIGNMENT message. This message is transmitted with four consecutive access bursts on the PACCH. The network evaluates the TA at the Rx of the access bursts. It provides the TA value in the PACKET POWER CTRL/TIMING ADVANCE message sent on the downlink PACCH.

During an uplink transfer, the BSS may initiate the establishment of a downlink TBF when it receives a downlink LLC PDU to transmit to the mobile. This can be done by sending either a PACKET DOWNLINK ASSIGNMENT or a PACKET TIMESLOT RECONFIGURE (see Figure 1.21) message on the PACCH.

Figure 1.21  Packet downlink establishment on the PACCH.

*RLC Principles*

The RLC layer provides a reliable radio link between the mobile and the network. The RLC data can be transmitted in RLC acknowledged mode or in RLC unacknowledged mode. The RLC layer performs the segmentation of the LLC frames that are received from the upper layer. One LLC frame is segmented into RLC data blocks. These are numbered and transmitted in sequence on the radio interface. Thanks to the numbering, the peer RLC is able to reorder the RLC data blocks and to perform the reassembly of the LLC frame. If some blocks have not been correctly decoded, the RLC peer entity can request their selective retransmission.

*Transmission Modes*  The RLC *automatic repeat request* (ARQ) functions support two modes of operation:

- RLC acknowledged mode;
- RLC unacknowledged mode.

RLC acknowledged mode is used to achieve high reliability in LLC PDU sending between the mobile and the network. The RLC ensures the selective retransmission of RLC data blocks that have not been correctly decoded by the receiver.

In RLC unacknowledged mode, the receiving entity does not request the retransmission of RLC data blocks that have not been correctly decoded. This mode is used for applications that are tolerant to error and request a constant throughput such as streaming application (video or audio streaming).

*Segmentation and Reassembly of LLC PDUs*  As the transmission unit size at the RLC layer is much lower than 100 bytes and the size of an LLC frame can be much larger, the segmentation mechanism allows the sharing of one LLC frame by several RLC data blocks. Depending on the CS used for the transmission through the air interface, the LLC frame is segmented into variable sized data units. Each data unit is encapsulated into one RLC data block that is numbered using the *block sequence number* (BSN) field of its RLC header. The BSN ranges from 0 to 127, and the RLC data blocks are numbered modulo 128.

The reassembly consists of the reordering of the RLC data blocks resulting from the BSN sequencing and of regenerating the LLC frame from the different data units that are contained in the RLC data blocks.

In RLC acknowledged mode this requires the correct reception of all the RLC data blocks that carry one part of the LLC frame. RLC unacknowledged mode, some RLC data blocks may not have been decoded during the transfer. The RLC data units not received have to be substituted with fill bits having the value 0.

*Transfer of RLC Data Blocks in Acknowledged RLC Mode*   The transfer of RLC data blocks in RLC acknowledged mode is controlled by a selective ARQ mechanism.

At the beginning of the TBF, the transmitter sends the RLC data blocks in sequence starting from BSN 0, BSN 1, and so on. However the maximum number of RLC data blocks that can be sent in sequence is controlled by a sliding window mechanism. The window size in GPRS is equal to 64. This means that when the mobile has sent BSN 63 it will not be allowed to transmit any new RLC data block (not previously transmitted) until the RLC data blocks with the lowest BSN have not been acknowledged. Once the RLC data block with BSN 0 has been acknowledged, the RLC data block with BSN 64 can be transmitted; once the RLC data block with BSN 1 has been acknowledged, the RLC data block with BSN 65 can be transmitted; the remaining blocks follow accordingly.

The window ensures that the gap, in term of block number, between the oldest unacknowledged block (the one that has the lowest BSN modulo 128) and the block that has been transmitted with the highest BSN modulo 128 is always lower than 64.

Figure 1.22 gives a scenario for uplink transfer.

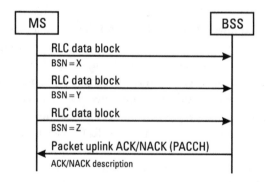

**Figure 1.22**   RLC data block transfer during uplink TBF.

*RLC Data Block Transfer During Uplink TBF* During uplink transfer, the mobile transmits one RLC data block in each uplink PDTCH instance allocated by the network. In order to acknowledge the RLC data blocks that have been correctly decoded at the BSS side, the network sends PACKET UPLINK ACK/NACK messages that contain a bitmap indicating, starting from a BSN value, all the blocks that have been correctly decoded or not. Each bit of the bitmap stands for the next in sequence BSN. A value 0 indicates that the block has not been correctly decoded, whereas 1 indicates the correct decoding of it.

Upon receipt of the PACKET UPLINK ACK/NACK message, the mobile starts the retransmission of the blocks that have not been acknowledged. Once they have been retransmitted, it resumes the transmission of new RLC data blocks as long as the RLC transmit is not stalled.

*RLC Data Block Transfer During Downlink TBF* During downlink transfer, the network controls the transmission of the RLC data blocks and the requests of acknowledgement from the mobile. Whenever needed, the network can request the transmission of a PACKET DOWNLINK ACK/NACK message from the mobile that indicates the RLC data blocks that have been correctly decoded or not.

The request of this message is performed by means of the *relative reserved block period* (RRBP) mechanism. The network sets a bit [*supplementary/polling* (S/P) bit] in the header of the RLC data block when it wants to receive an acknowledgement message. It indicates the number of frames that have to elapse between the reception of the data block containing the polling indication and the beginning of the transmission of the acknowledgement message. The PACKET DOWNLINK ACK/NACK message is sent on the uplink PDCH associated to the downlink one on which has been received the polling indication. Figure 1.23 shows a scenario for downlink transfer.

*Transfer of RLC Data Blocks in RLC Unacknowledged Mode* In RLC unacknowledged mode, the sending side transmits the RLC data blocks in sequence. The blocks are numbered so that the receiver is notable to detect any decoded data blocks. The data blocks that are not decoded by the receiver are not retransmitted. On the receiver side when all the RLC data blocks belonging in one LLC frame have been received, it reassembles the LLC frame.

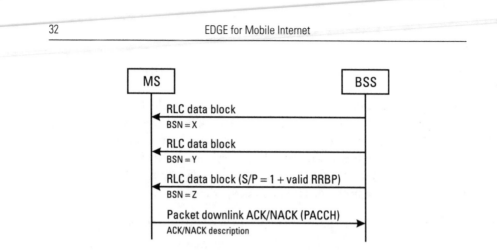

**Figure 1.23** RLC data block transfer during downlink TBF.

*RR Release*

*Release of Uplink TBF* The release of the uplink RRs is controlled by a count-down procedure. When the mobile starts to send the last 16 radio blocks, it triggers a countdown procedure and indicates the *countdown value* (CV) in

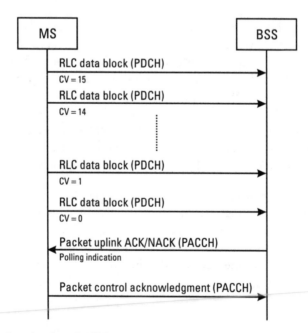

**Figure 1.24** Procedure for uplink TBF release.

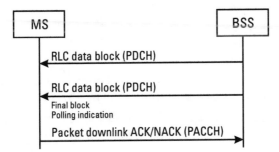

**Figure 1.25** Procedure for downlink TBF release.

the header of the RLC blocks. When the network receives the last RLC data block (CV equal to 0) and if all the blocks have been correctly decoded, it sends a PACKET UPLINK ACK/NACK message that indicates the release of the RRs. An acknowledgment is requested to be sure that the mobile has received the release order. The acknowledgment is requested by means of the RRBP procedure described in the section "Transfer of RLC Data Blocks in Acknowledged RLC Mode."

Upon receipt of the PACKET UPLINK ACK/NACK message, the mobile sends a PACKET CONTROL ACKNOWLEDGEMENT message on PACCH and releases the TBF. The release procedure is illustrated in Figure 1.24.

*Release of Downlink TBF*   During a downlink transfer, when the BSS sends the last RLC data block belonging to the TBF it indicates within its header that it is the final block. The network requests the sending of a PACKET DOWNLINK ACK/NACK message that confirms the release of the TBF by introducing a polling indication in the final block. Upon receipt of the PACKET DOWNLINK ACK/NACK message, the resources are released if all the blocks are positively acknowledged. The procedure is illustrated in Figure 1.25.

### 1.3.3   Cell Reselection

Unlike GSM circuit-switched services, there is no handover process in GPRS. This means that the MS cannot perform a seamless cell change while transmitting or receiving on a PDTCH. Instead, cell reselection is possible. The principle is similar to GSM reselection.

There are three selection modes, as follows:

- *Mode NC0.* The GPRS mobile performs autonomous cell reselection without sending measurement reports to the network,
- *Mode NC1.* The GPRS mobile performs autonomous cell reselection and periodically sends measurement reports to the network,
- *Mode NC2.* The network controls the cell reselection. The mobile sends measurement reports to the network.

The cell reselection can be either controlled by the network or autonomously performed by the mobile. When the mobile performs autonomous cell reselection, it chooses a new cell and triggers a cell reselection on its own. Cell reselection is based on measurements performed by the mobile. The network can order a report of these measurements periodically or not.

**Table 1.5**
Mode of Reselection and Criteria for Cell Reselection

| NETWORK_ CONTROL_ ORDER Value | GMM State of the MS | Mode of Cell Reselection | Criteria If PBCCH Exists | Criteria If PBCCH Does Not Exist |
|---|---|---|---|---|
| NC0 | Standby/Ready | Autonomous cell reselection | C'1, C31, C32 | C1, C2* |
| NC1 | Standby | Autonomous cell reselection | C'1, C31, C32 | C1, C2* |
|  | Ready | Autonomous cell reselection with measurement reports | C'1, C31, C32 | C1, C2* |
| NC2 | Standby | Autonomous cell reselection | C'1, C31, C32 | C1, C2* |
|  | Ready | Network-controlled cell reselection with measurement reports | — | — |

*Except if the GPRS cell reselection parameters are sent to the MS in an RLC/MAC control message.

The GPRS cell reselection mode for a GPRS attached MS is given by the network control mode (NETWORK_CONTROL_ORDER parameter) that is broadcast on the BCCH or PBCCH. The mobile's behavior is determined by both its GMM state and the network control mode.

When the PBCCH is present in the serving cell, the network broadcasts both the neighbor cell and serving cell parameters that are used for cell reselection on this channel. When there is no PBCCH in the serving cell, these parameters are broadcast by the network on BCCH. The serving cell parameters used for cell reselection are broadcast on the serving BCCH, whereas the neighbor cell parameters are broadcast on the neighbor BCCHs.

Two criteria are defined for autonomous cell reselection—one that is the same as in GSM, and a new one specific to GPRS. The description of these criteria exceeds the scope of this book. Refer to [1] for further detail. Table 1.5 summarizes the different reselection modes and the criteria used.

## 1.4 GPRS Mobility Management

The continuity of packet services is ensured in the network owing to the management of GPRS mobility. The network needs to know the *location area* (LA) of GPRS subscribers so that the network can page them in case of incoming packet-switched calls. Thus the GMM procedures make it possible to track GPRS subscribers when they move from one LA to another one area within a given PLMN.

A subscriber has access to GPRS services only if it has already been IMSI attached for GPRS services. A subscriber does not have access to a GPRS service any more when it is GPRS detached. An LA dedicated to GPRS services is called a *routing area* (RA). It is made up of a subset of cells defined by the operator of the PLMN network; an RA is identified by a *routing area identifier* (RAI), whereas a cell is identified by a *cell identifier* (CI). An LA dedicated to non-GPRS services is made up of one or several routing areas. An RA defines a paging area for incoming packet-switched calls. Thus a GPRS MS is paged in every cell belonging to the RA in case of an incoming packet-switched call if this one is not located at cell level.

The SGSN of an MS handles a GPRS mobility context related to it. Information such as IMSI, P-TMSI, RAI, CI, as well as the GMM state is stored in the mobility context by the SGSN as well as by the MS in the SIM card.

### 1.4.1   GMM States

Three different GMM activities—IDLE, STANDBY, READY—related to a GPRS subscriber are defined to characterize the GMM activity. This information is updated in the mobility context at the MS and SGSN sides. The behavior of the MS on the radio interface depends on the GMM state. It follows that the GMM state is known by the RRM layer of the MS and by the BSS.

In GMM IDLE state, a subscriber does not have access to GPRS services. No GPRS mobility context is established between the MS and the SGSN. No GMM procedures are performed in this state.

In the GMM STANDBY state, a subscriber can have access to GPRS services. A subscriber enters into STANDBY state from READY state either upon expiry of the READY timer or upon explicit request from the network. A GPRS mobility context is created between the MS and SGSN; the loca-

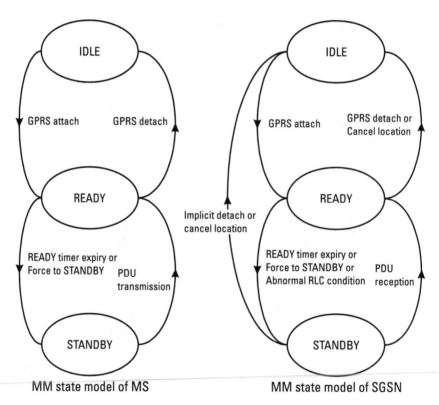

**Figure 1.26**   Global states of GPRS mobility. (*From:* [3]. © ETSI 2001.)

tion of the subscriber is known by the network at the RA level. In this state, the MS may receive paging for GPRS data or packet-switched signaling as well as paging for circuit-switched calls. GPRS cell selection, GPRS reselection, and GPRS location procedures may be also performed in this state.

In GMM READY state, a subscriber may send or receive GPRS data or packet-switched signaling. A subscriber enters into READY state either from IDLE state at the end of a successful IMSI attach procedure for GPRS service or from STANDBY state each time a packet is sent to the network. A READY timer is reinitialized by the MS and SGSN for each packet sent by the MS. In READY state the subscriber is located by the network at cell level. GPRS paging is never initiated by the network in READY state to know the MS location. In READY state, GPRS cell selection, GPRS reselection, radio link measurement reporting, GPRS location procedures, and notification of cell change may be performed. Figure 1.26 shows the transitions between the three GMM states.

### 1.4.2 GPRS MS Classes

Three classes of GPRS mobiles have been defined; they are as follows:

- *Class A.* A class A mobile is simultaneously attached to non-GPRS and GPRS services. It is able to support simultaneously one communication in circuit-switched mode and another one in packet-switched mode.

- *Class B.* A class B mobile is simultaneously attached to non-GPRS and GPRS services. It cannot support simultaneously both types of service in communication, but it is able to detect an incoming circuit-switched call or an incoming packet-switched call during idle mode.

- *Class C.* A class C mobile is only attached to non-GPRS services or to GPRS services. The use is alternated either manually by the user or automatically by the application.

### 1.4.3 Mobility Procedures

1.4.3.1 Paging

An MS may be paged by the network for circuit-switched and packet-switched services. Coordination of paging for circuit-switched and packet-switched services may be provided by the network so that the MS can receive the paging for non-GPRS services and GPRS services on the same logical channels. The coordination takes place only if the Gs interface between SGSN and MSC/VLR is present.

Three *network modes of operation* (NMOs) have been defined; they are as follows:

- *Mode I.* The network sends paging messages on the same logical channels for non-GPRS and GPRS services since paging coordination is supported (i.e., on PCCCH paging channels if allocated in the cell, and on CCCH paging channels otherwise).

- *Mode II.* The network sends paging messages for non-GPRS and GPRS services on CCCH paging channels.

- *Mode III.* The network sends paging messages for non-GPRS services on the CCCH paging channels and paging messages for GPRS services on the PCCCH paging channels if they exist or on the CCCH paging channels otherwise.

### 1.4.3.2   GPRS Attach

When an MS needs to access GPRS services, it performs an IMSI attach for GPRS services to signal its presence to the network. During this procedure, the subscriber provides its identity either with a temporary *packet temporary MS identity* (P-TMSI) identifier or with an *international mobile subscriber identity* (IMSI) identifier.

Two types of GPRS attach procedures are defined here:

- *Normal GPRS attach.* This procedure is used by the MS to be IMSI attached only for GPRS services.

- *Combined attach procedure.* This procedure is used by a class A or class B MS to be IMSI attached for non-GPRS and GPRS services in a cell operating in mode I.

A *mobility management* (MM) context is created between the MS and the SGSN at the end of the procedure.

### 1.4.3.3   GPRS Detach

An IMSI detach procedure for GPRS services is initiated either by the MS or by the SGSN to release the MS access to GPRS services. This procedure allows the network to avoid wasting RRs in case of incoming packet-switched calls when the MS is detached for GPRS services.

Two types of GPRS detach procedures are defined as follows:

- *Normal GPRS detach.* This is used to IMSI detach only for GPRS services.

- *Combined detach procedure.* This procedure is used to IMSI detach for circuit-switched and GPRS services a class A or class B MS in a cell operating in mode I.

The MM context between the MS and the SGSN is removed at the end of the procedure.

### 1.4.3.4   Security Functions

The authentication procedure is used by the network to identify and authenticate the subscriber. It makes it possible to protect the radio link from unauthorized calls afterwards. Each GPRS authentication includes a triplet:

- Specific ciphering key Ki known by the MS and network;
- Random number provided by the HLR/AUC and sent to the MS;
- SRES, the response of the authentication request.

SRES is a number calculated by the HLR/AUC and MS from the algorithm A3 with a key (Ki) that is specific to the GPRS subscriber. The ciphering key Kc for a GPRS subscriber is also calculated by the HLR/AUC and MS from the algorithm A8 and key Ki.

The network guarantees confidentiality of user identity when a subscriber has access to GPRS RRs. Confidentiality of user identity is ensured by the P-TMSI identifier. On the radio interface, a *temporary logical link identity* (TLLI) identifies a GPRS subscriber within a RA; it is deduced from the P-TMSI. The relation between the TLLI and the IMSI is known only by the MS and the SGSN.

An SGSN may request the identity of the mobile. It can thus verify the identifier of the *international mobile equipment identity* (IMEI) returned by the MS and compare it with the identifier stored in the EIR.

The network guarantees confidentiality of the call by ciphering it. Data ciphering for GPRS is done at the LLC layer level between the MS and SGSN.

#### 1.4.3.5  Location Updating Procedures

A location procedure is always initiated by the MS. It may occur each time that an MS camps on a new cell for better radio conditions. The type of location procedure also depends on the GMM state, the NMO, and the class of the MS. Thus, an MS analyzes the CI, the RAI, and the *location area identifier* (LAI) of the new cell.

An MS performs a *cell update* (CU) procedure when it camps in a new cell within its current RA only in GMM READY state, since the location of subscriber in this GMM state is known by the network at the cell level.

An MS performs an RA update procedure in order to update MM context between the MS and SGSN when it camps in a new cell belonging to a new RA. This procedure may also occur at the expiry of a periodic timer in order to check the presence of the subscriber in an RA.

Four types of RA update procedures are defined; they are as follows:

- *Normal RA update*—performed by a class C MS or by a class A or B MS upon detection of a new routing area in a cell operating in mode II or III;

- *Periodic RA*—performed by any GPRS MS upon expiry of a timer;

- *Combined RA and LA update*—performed by a class A or B MS upon detection of a new LA in a cell operating in mode I;

- *Combined RA with IMSI attach*—performed by a class A or class B MS already GPRS attached in a cell operating in mode I in order to be IMSI attached for non-GPRS services.

## 1.5  PDP Context Management

A PDP context makes it possible to characterize an access to an external packet-switching network. It contains information such as *access point name* (APN), which is the reference of GGSN; *LLC service access point identifier* (LLC SAPI), which identifies the *service access point* (SAP) used for GPRS data transfer at the LLC layer; *network service access point identifier* (NSAPI), which identifies the SAP used for GPRS data transfer at the SNDCP layer; the requested QoS; and the type of packet-switched network. A PDP context is identified by an MS PDP address within the MS, SGSN, and GGSN entities. Several PDP contexts can be activated simultaneously in a given MS and are identified by several MS PDP addresses.

Before GPRS data transfer within an external packet-switching network, a PDP context must be in an active state. A PDP context activation procedure initiated either by the MS or by the network is used to create a PDP context. A PDP context is deactivated by a PDP context deactivation procedure initiated by the MS or by the network (SGSN or GGSN). A PDP context is modified by a PDP context modification procedure initiated either by the MS or by the network in order to change some parameter values such as requested QoS. The SM protocol and *GPRS tunneling protocol* (GTP) handles the PDP context procedures respectively between the MS and the SGSN and between the SGSN and the GGSN.

The MS PDP address that is an IP address can be assigned statically or dynamically during the PDP context activation procedure. A PDP address, assigned statically at the time of subscription, is called a static PDP address. A PDP address, assigned dynamically either by the GGSN or by the PDN operator, is called a dynamic PDP address.

A concept of secondary PDP context has been defined in Release 99 of 3GPP; it allows the reuse of the PDP address, the APN, and other information from an already active PDP context with a different QoS profile. This principle is useful for multimedia applications where each medium type requires specific transport characteristics and needs to be mapped into a specific PDP context. A filtering mechanism is used by the GGSN to route the IP packets from the external data packet network toward the appropriate medium. This mechanism is based on a *traffic flow template* (TFT) that is defined by a set of packet filters. Each packet filter contains a list of attributes, each attribute being deduced from IPv4 or IPv6 headers.

## 1.6 GPRS Backbone Network

A GPRS backbone network is made up of GSNs. User packet and signaling are conveyed across the Gn/Gp interface in the GPRS backbone. The gray boxes in Figure 1.27 delimit the GPRS backbone network.

The GTP layer provides services for carrying user data packets and signaling between the GPRS support nodes in the GPRS backbone network. Packets from MS or the external data packet network are encapsulated by *GPRS tunneling protocol for the user plane* (GTP-U) in *GTP-U PDUs* (G-PDUs) and are tunneled through the GPRS backbone network. Signaling messages between GSNs are also tunneled by the *GTP for the control plane* (GTP-C).

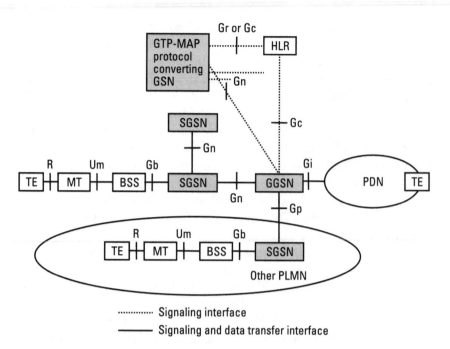

**Figure 1.27**  Architecture of GPRS backbone network. (*From:* [4]. © ETSI 2001.)

A GTP tunnel is a two-way PTP means to forward packets between two GSNs. It is identified in each GSN node by a *tunnel endpoint identifier* (TEID), an IP address, and a UDP port number. UDP/IP are the backbone network protocols used for user data routing and control signaling. The IP address and UDP port number define a UDP/IP path that is connectionless between two GSNs. Thus, for a UDP/IP path the IP source address is the IP address of the source GSN, whereas the IP destination source is the IP address of the destination GSN. The TEID identifies the tunnel endpoint in the receiving GTP protocol entity and enables the multiplexing of GTP tunnels between a given GSN-GSN pair on a UDP/IP path.

A GTP-U tunnel is a tunnel in the user plane defined for a PDP context in the GSNs and is used to route user data between the MS and an external data packet network. A GTP-U tunnel in control plane is a tunnel defined for all PDP contexts with the same PDP address and APN. GTP tunnels are created, modified, and deleted with the tunnel management procedures.

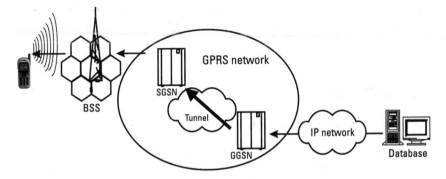

**Figure 1.28** Tunneling mechanism in user plane for IP packet sending toward MS.

## 1.6.1 GTP-U

The GTP-U conveys an IP datagram between MS and the external PDN in the GPRS backbone network. The IP datagram tunneled in the GTP-U tunnel is called a T-PDU. A GTP header with the TEID field is added to the T-PDU in order to constitute a G-PDU. In this manner T-PDUs may be multiplexed between a given GSN-GSN pair on a UDP/IP path by the GTP layer. Figure 1.28 shows a tunneling mechanism in the user plane for IP packet sending toward MS.

## 1.6.2 GTP-C

The GTP-C tunnels signaling messages in the GPRS backbone network. A GTP header with the TEID field is added to GTP signaling message to constitute a GTP-C PDU, which is sent in a UDP/IP path.

The GTP-C enables performance of several procedures through the GPRS backbone network, (e.g., path management, tunnel management, location management, and mobility management). The path management procedure is used to find out if the peer GSN is alive. The tunnel management procedures are used to create, update, and delete GTP tunnels in the GPRS backbone network. The location management procedure is used to transport location messages between a GGSN, which does not have an SS7 MAP interface, and a GTP-MAP protocol-converting GSN in the GPRS backbone network; this procedure may occur when the network requests the activation of a PDP context. The mobility management procedure is used to update the MM and PDP context information in a new SGSN from an old SGSN; this procedure may occur during GPRS attach or inter-SGSN routing area update procedures.

## 1.7 CAMEL for GPRS

### 1.7.1 Mobile Market Evolution

Up to now, the mobile market has been driven by voice. Since the mobile voice market is already mature in numerous countries, the operators need new services to increase the *average revenue per user* (ARPU). The new data services have to allow the operators to generate these new revenues. One way is to extend the prepaid services for GPRS. The prepaid is a recognized payment mode with 60% of subscribers in Europe. The *customized applications for mobile network enhanced logic* (CAMEL) feature provides mechanisms to support specific operator services for mobiles when roaming outside the HPLMN, such as prepaid services. CAMEL phase 3, introduced in Release 99 of the 3GPP recommendations, offers capabilities to support GPRS roaming for prepaid subscribers, whereas CAMEL 1 and 2 were intended for voice prepaid subscribers.

CAMEL phase 3 is a key enabler of the GPRS prepaid market since it supports billing. For charging, the CAMEL phase 3 takes into account several parameters such as volume, duration, QoS, and location.

### 1.7.2 Architecture for GPRS CAMEL Services

CAMEL reuses the concept of *intelligent network* (IN) which separates functions common to all applications (e.g., call control) and specific functions related to one application or one service. Common functions are managed by switching centers, and specific functions are integrated in a *service control point* (SCP). In the IN terminology, the switching centers become *service switching point* (SSP). The SCPs are equipment nodes able to exchange signaling information with SSPs. The purpose of the IN architecture is to allow the smooth introduction of new services in the network, because it is easier to add a new SCP dedicated for each new service without updating the release of switching centers.

In CAMEL, an SCP is called a *CAMEL Service Environment* (CSE). In a GPRS network an SGSN contains common functions related to SSP and the SAP.

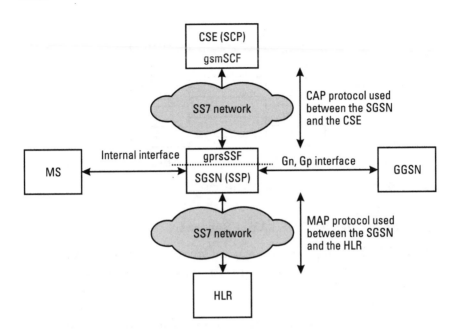

**Figure 1.29** Architecture for GPRS CAMEL services.

The specific services provided by an operator are called *operator-specific services* (OSS). The functional entity in the CSE, which contains the CAMEL service logic to implement OSS, is called a *GSM service control function* (gsmSCF). The functional entity in the SGSN that interfaces with the gsmSCF is called a *GPRS service switching function* (gprsSSF). Figure 1.29 illustrates the architecture for CAMEL services.

The IN networks are based on SS7 networks. The MTP, SCCP, and TCAP protocol layers are used in IN networks between the SSP and SCP. For GPRS CAMEL services, a *CAMEL application part* (CAP), an application protocol is used between the SGSN (SSP) and the CSE (SCP); it was already used for GSM CAMEL services. This protocol is close to the *IN application part* (INAP) defined for fixed networks. Figure 1.30 shows the protocol stack used between the SGSN and the CSE.

The GPRS CAMEL services related to a subscriber are identified by the *GPRS-CAMEL subscription information* (GPRS-CSI). The GPRS-CSI contains information related to the OSS of the subscriber, the GPRS CAMEL service logic that is to be applied by the gsmSCF, and the CSE address (E.164 number) to be used for gsmSCF access. This latter is stored in the HLR.

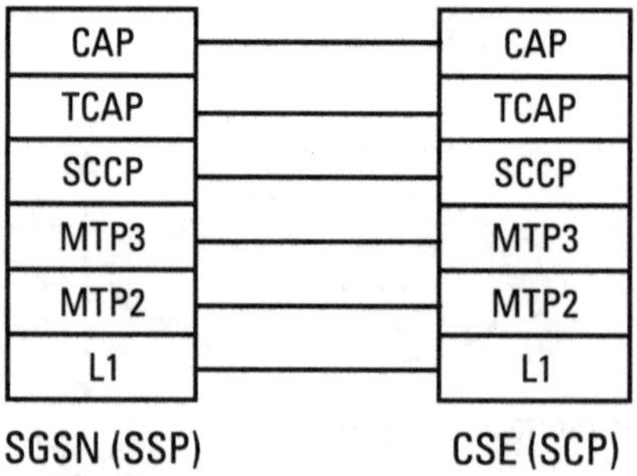

**Figure 1.30**  Protocol stack used between the SGSN and CSE.

### 1.7.3   Procedures for GPRS CAMEL Services

The GPRS CAMEL procedures may be invoked at the time of the GPRS attachment procedure, the PDP context activation procedure, or the routing area update procedure in order to monitor or modify the handling of these procedures.

In order to invoke GPRS CAMEL procedures, *detection points* (DPs) are used to detect GPRS events that are to be notified to the gsmSCF. This latter can potentially influence the GPRS, the session, and the PDP contexts.

Two types of CAMEL relationships are defined; they are as follows:

- Monitor relationship. The gsmSCF receives GPRS events from gprsSSF but does not send any instructions to the gprsSSF; the GPRS session or PDP context procedure is not suspended.

- Control relationship. The gsmSCF is able to influence the GPRS session by sending some instructions to the gprsSSF; the GPRS session or PDP context activation procedures are suspended in the GPRS network upon GPRS event detection as long as the gprsSSF receives instructions from the gsmSCF.

Figure 1.31 is an example scenario where the initial GPRS event occurs during a GPRS attach procedure. The GPRS attach procedure is suspended by the SGSN until the gprsSSF receives instructions from CSE. The instruction command may be a request to control the charging of a GPRS session.

Figure 1.32 is an example scenario where the GPRS event occurs during a PDP context establishment procedure. The PDP context establishment procedure is suspended by the SGSN until the gprsSSF receives instructions from CSE.

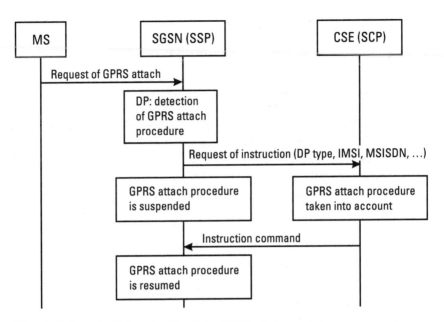

**Figure 1.31** Example of information flow during a GPRS attach procedure.

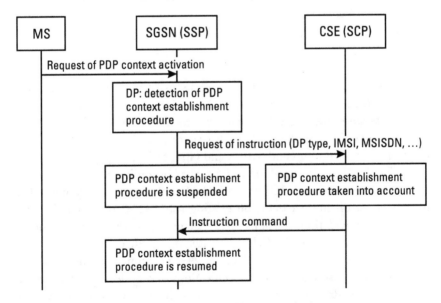

**Figure 1.32** Example of information flow during a PDP context activation.

## 1.8    Organization of the 3GPP

The GPRS specifications are part of the GSM recommendations since the Release 97.

The responsibility for GSM specifications was carried by ETSI up to the end of 1999 and was transferred to the *Third Generation Partnership Project* (3GPP) during the year 2000. This world organization was created to produce the third-generation mobile system specifications and technical reports, including evolved radio access technologies such as GPRS and *enhanced data rates for global evolution* (EDGE).

The 3GPP consists of the following *technical specification groups* (TSG):

- *Service Architecture* (TSG SA)—dealing with service, architecture, security, speech coding aspects;
- *Radio Access Network* (TSG RAN)—dealing with UTRA radio access technologies;
- *Core Network* (TSG CN)—dealing with core network specifications;
- *Terminal* (TSG T)—dealing with applications, tests for 3G mobiles, USIM card;
- *GSM EDGE Radio Access Network* (TSG GERAN)—dealing with GSM radio interface, A and Gb interfaces.

Thus, GSM evolutions as GPRS are discussed in all TSGs except TSG RAN which deals exclusively with UTRAN access technologies such as *frequency division duplex* (FDD), *time division duplex* (TDD), and *code division multiple access* (CDMA2000). TSG GERAN deals exclusively with the GSM radio interface evolutions and with A and Gb interfaces.

The 3GPP recommendations are ranked according to a version reference. Each new version of 3GPP recommendations contains a list of new features or a list of improvements on existing features. At the beginning, the GSM recommendations versions were referenced in the following order: Phase 1, Phase 2, Release 96, Release 97, Release 98, Release 99. As the reference year for the new version of phase 2++ recommendations no longer matched the release year of these specifications, it was decided that the versions following Release 99 will be referenced from now on with a version number, Release 4 being the first new version reference.

The GSM recommendations are organized up to Release 99 in the following series:

- 01 series—General;
- 02 series—Service aspects;
- 03 series—Network aspects;
- 04 series—MS–BS interface and protocols;
- 05 series—Physical layer on the radio path;
- 06 series—Speech coding specification;
- 07 series—Terminal adaptors for MSs;
- 08 series—BS–MSC interface;
- 09 series—Network interworking;
- 11 series—Equipment and type approval specification;
- 12 series—Operation and maintenance.

Each of the series contains a list of specifications identified by numbers. A given specification is therefore defined by its series number, followed by a recommendation number. For example, the 05.03 specification belongs to the physical layer on the radio path and deals with channel-coding issues.

## References

[1] Seurre, E., P. Savelli, and P. J. Pietri, *GPRS for Mobile Internet*, Norwood, MA: Artech House, 2003.

[2] 3GPP TS 03.64 Overall Description of the GPRS Radio; Stage 2 (GPRS).

[3] 3GPP TS 23.060 Service Description, Stage 2 (GPRS).

[4] 3GPP TS 29.060 GPRS Tunneling Protocol (GTP) Across the Gn and Gp Interface.

## Selected Bibliography

3GPP TS 23.078 Customized Applications for Mobile Network Enhance Logic (CAMEL) Phase 3; Stage 2 (R99).

3GPP IS 25.061 Interworking Between the Public Land Mobile Network (PLMN) Supporting Packet Based Services and Packet Data Networks (PDN) (R99).

# 2

# Introduction to EDGE

This chapter is the first of a series dedicated to EDGE. It introduces the different EDGE concepts from a global point of view, explaining how they have been introduced into the packet-switched and circuit-switched networks. However, the goal of these chapters is to focus only on EDGE for the packet-switched domain.

The various EDGE concepts are briefly introduced in the first section of this chapter. The second section is dedicated to the services that are provided by EDGE. The last section gives an overview of the *enhanced general packet radio service* (EGPRS) concept. It delineates the basic principles of EDGE together with the changes as compared with GPRS. It is necessary to understand these principles in order to have a global vision of how EGPRS is working and to be able to understand the following chapters, which will discuss this concept in greater detail.

## 2.1    What Is EDGE?

EDGE is a global radio–based high-speed mobile data standard that can be introduced into GSM/GPRS and IS-136 [packet mode for *digital advanced mobile phone system* (D-AMPS)] networks. EDGE allows data transmission speeds up to 384 Kbps in packet-switched mode; these throughputs are required to support multimedia services. This is achieved within the same GSM bandwidth and existing 800-, 900-, 1800-, and 1900-MHz frequency bands.

The idea behind EDGE is to increase the data rate that can be achieved with the 200-kHz GSM radio carrier by changing the type of modulation used while still working with existing GSM and GPRS network nodes. The new modulation that is introduced is the *eight-state phase-shift keying* (8-PSK). The basic concept constraint was to have the smallest possible impact on the core networks.

EDGE is considered in Europe as a *2.5 generation* (2.5G) standard that is seen as a transition from 2G to 3G (second generation and third generation of mobile networks). No new operator licenses are needed for EDGE. Since this feature reuses the existing spectrum, it represents a low-cost solution for operators that want to provide multimedia services on their GSM/GPRS networks. However, in some countries such as the United States, for operators that do not have *Universal Mobile Telecommunications System* (UMTS) licenses, EDGE can provide multimedia services that will be brought by 3G networks and that cannot be supported by the GPRS system. That is why EDGE may also be viewed as a 3G standard. Since EDGE can be considered as a low-cost solution, it can also be used by operators that already have a GPRS network and a UMTS license to provide 3G services within areas where a UMTS coverage would not be cost-effective.

EDGE amounts to a global evolution of the circuit-switched GSM network, the packet-switched GPRS network, and the D-AMPS network since it has been introduced in all these technologies (see Figure 2.1).

EDGE was introduced on top of the *high-speed circuit-switched data* (HSCSD) service that is used to transfer data in circuit-switched mode on several time slots. The evolution of this service with the introduction of EDGE is called *enhanced circuit-switched data* (ECSD). ECSD supports the current GSM rates (2.4 Kbps, 4.8 Kbps, 9.6 Kbps and 14.4 Kbps) and new CSs that combined with the new modulation, allow the following data rates: 28.8 Kbps, 32.0 Kbps, and 43.2 Kbps per time slot. Thus, a multi-

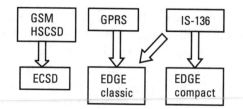

**Figure 2.1**   Convergence of the different standards for EDGE.

slot mobile supporting ECSD on four time slots can reach a throughput up to 172.8 Kbps. ECSD uses dynamic link adaptation to adapt the throughput to the radio conditions under which the mobile is operating.

The evolution of GPRS toward EDGE is called EGPRS. It is sometimes also called EDGE Classic. EGPRS is based on the same network architecture as GPRS. It allows throughputs up to 475 Kbps for a receiver supporting Rx on eight time slots. The interface most affected is the radio interface, owing to the introduction of the new radio modulation. The major impacts are located on the BSS part of the network and the MS.

EDGE is also an evolution of the IS-136 U.S. standard. The standardization of EDGE allowed U.S. operators and manufacturers to have a global standardized solution for packet-switched network. U.S. operators already operating D-AMPS networks wanted to introduce 2.5G (or 3G) services in their current networks. However, these operators wanted to introduce EDGE services within the same frequency band as D-AMPS—the 800-MHz band; but this band, which is split into 30-kHz radio carriers, is already used for speech services. The introduction of EDGE requires the release of some 30-kHz carriers in order to provide the 200-kHz carriers necessary to operate EDGE. These carriers once allocated to EDGE cannot be used for speech services (this is not the case for GSM/EDGE). It is possible to release only a small part of the 800-MHz spectrum for the EDGE services in order not to damage the network's speech capacity. So it was decided to standardize a spectrum-efficient version of EGPRS. It will support 384-Kbps mandated packet data rates but will require only minimal spectral clearing and therefore could work for network operators with limited spectrum allocations. This concept is called EDGE Compact. Thus EDGE Compact is an adaptation of EGPRS that introduces special features for operating in a spectrum-limited network. EDGE Compact is limited to the packet-switched mode; it does not concern the circuit-switched mode.

The cohabitation between D-AMPS and EDGE systems has required the introduction of new radio requirements in the GSM standard (in order that the two systems can live together). As a result, EDGE, which previously stood for *enhanced data rates for GSM evolution,* now stands for *enhanced data rates for global evolution.* The EDGE Compact standard was jointly standardized by the *Universal Wireless Communications Corporation* (UWCC), the ETSI, and the 3GPP.

However, it seems that the major wireless operators in the United States, where EDGE Compact was supposed to be deployed, have decided to remove their D-AMPS network and replace it with the GSM/GPRS/EDGE technology. This is the reason why there is a high probability that EDGE Compact will never be deployed in the field. In the following sections, the EGPRS concept is described. ECSD and EDGE Compact are not discussed in this book.

## 2.2    EGPRS Services

### 2.2.1    EGPRS General Characteristics

EGPRS is a direct evolution of GPRS. It reuses the same concepts and is based on exactly the same architecture as GPRS. The introduction of EGPRS has no impact on the GPRS core network. The main modifications are linked to the radio interface.

The EGPRS concept aims at enabling data transmission with higher bit rates than GPRS. Basically, EGPRS relies on a new modulation scheme and new CSs for the air interface, making it possible to optimize the data throughput with respect to radio propagation conditions. Nine *modulation and coding schemes* (MCSs) are proposed for enhanced packet data communications, providing raw RLC data rates ranging from 8.8 Kbps (minimum value per time slot under the worst radio propagation conditions) up to 59.2 Kbps (maximum value achievable per time slot under the best radio propagation conditions). Data rates above 17.6 Kbps require that 8-PSK modulation be used on the air instead of the regular GMSK modulation. Table 2.1 gives the throughputs associated with the various MCSs.

On top of the same services as GPRS, EGPRS provides new ones because of higher bit rates. Moreover, it basically offers twice the capacity of a GPRS network. Indeed, although the bit rate of the modulation is increased by a factor 3 with the new modulation, allowing a maximum throughput that is three times higher, the capacity of the network is not multiplied by 3. This is due to the *carrier-to-interference ratio* (C/I) variation within the network. Depending on the MS position, more or less channel coding will be necessary for an optimized transmission, leading to an average throughput lower than the maximum one.

EGPRS provides the most cost-effective means to provide 3G services within the existing spectrum. It allows operators to deliver new 3G services by upgrading their existing GSM/GPRS wireless infrastructures. The hardware modifications in the network are limited to the addition of new EDGE transceiver units in each cell. These units can be added in complement to the existing devices that have already been deployed.

**Table 2.1**
Throughput Associated with Modulation and Coding Schemes

| Modulation and Coding Scheme | Modulation | Maximum Throughput (Kbps) |
| --- | --- | --- |
| MCS-9 | 8-PSK | 59.2 |
| MCS-8 | 8-PSK | 54.4 |
| MCS-7 | 8-PSK | 44.8 |
| MCS-6 | 8-PSK | 29.6 |
| MCS-5 | 8-PSK | 22.4 |
| MCS-4 | GMSK | 17.6 |
| MCS-3 | GMSK | 14.8 |
| MCS-2 | GMSK | 11.2 |
| MCS-1 | GMSK | 8.8 |

EGPRS allows the support of data and multimedia services and applications—services that cannot be supported with GPRS (or only with very poor quality). Radio and video broadcasts via wireless phones will then be possible.

## 2.2.2 EGPRS MS Capabilities

An EGPRS MS is characterized by different parameters that give information on its capabilities. The consequence of the introduction of a new modulation is the implementation of a totally new part in layer 1 of the mobile dedicated to it. This leads to special characteristics typical of EGPRS MSs. This modulation has brought about new constraints in the design of layer 1 on the RF part as well as on the baseband part.

The introduction of EGPRS capabilities within the MS has deep impacts on the baseband part. The introduction of 8-PSK with a higher spectral efficiency requires new and more complex algorithms for the equalization part as well as for the coding and decoding part. The level of complexity is increasing greatly as compared with the GMSK modulation. The execution of the same number of actions during one TDMA frame (encoding, equalization, decoding) requires much more processing. A mobile that can perform $x$ receptions during one frame in GPRS mode may only perform $y$ receptions in EGPRS mode with $y \leq x$.

As a result, an EDGE MS will be characterized by two multislot classes: one corresponding to its pure GPRS multislot class and another to its pure EGPRS multislot class. The EGPRS multislot class of the mobile is defined by the same parameters as the GPRS one: maximum number of RX time slots, maximum number of TX time slots, and maximum number of RX + TX time slots per TDMA frames (see Section 1.3.1.3 in Chapter 1). The MS shall support the EGPRS multislot class in both EGPRS and GPRS mode. For example, the multislot class of the mobile can be 4 RXs + 1 TXs in pure GPRS mode and 2 RXs + 1 TX in pure EGPRS mode. The GMSK modulation signal that is used for GPRS transfer has the characteristic of having a constant amplitude. This reduces the transmitter design constraint. But the EDGE 8-PSK is a phase modulation with amplitude variations due to the transitions between the different constellation symbols. This amplitude variation brings a higher level of complexity to the implementation of the transmitter, in particular in the design of the amplifier, which requires a higher level of linearity. This has a significant impact on *power amplifier* (PA) efficiency (see Section 3.4.3 in Chapter 3). In order to reduce the complexity of the mobile and to provide EGPRS-capable MSs as early as possible on the market, two classes of MSs have been defined.

The first class supports 8-PSK and GMSK modulations in the downlink direction reception but is limited to GMSK modulation in uplink. This means that a mobile from this class supports MCS-1 to MCS-9 in reception but is capable only of MCS-1 to MCS-4 in uplink (of course the GPRS CSs are also supported). The second class supports 8-PSK in both uplink and downlink directions. This kind of mobile supports all the MCSs in Rx and Tx.

The 8-PSK modulation is not necessarily needed in uplink because of the asymmetry in the throughput between uplink and downlink for some services. In fact most of the services such as video broadcasts and Web browsing require a high throughput in downlink, whereas the uplink is used only for the transmission of signaling and commands. This has also been a reason for not imposing support of the 8-PSK modulation in uplink on the mobile side.

Owing to the implementation constraints of the PA, it will be difficult to have the same output power when transmitting using GMSK or 8-PSK modulation. Three different power classes for 8-PSK modulation have been defined. A mobile is then characterized by its GMSK power class and its 8-PSK power class. The nominal maximum output power can be different (see Section 3.2.1 in Chapter 3).

## 2.3   EGPRS General Principles

EGPRS Classic, which is the direct evolution of GPRS, reuses to the maximum all the basic concepts of the GPRS system. The great advantage of EGPRS compared with GPRS is in the way the radio transmission is managed. This improvement is brought by the use of a new modulation and the introduction of new optimized coding families that allow a very efficient link quality control mechanism. The RLC protocol has been slightly improved so that it provides sufficient efficiency for the transmission of higher throughput. This chapter gives the basic principles of the EGPRS system and explains the main improvements compared with GPRS.

### 2.3.1   EGPRS Basis

The EGPRS system relies on the same architecture as GPRS. The part of the network most affected is the BSS. In spite of the changes to the radio interface, the same layer structure for signaling and data as for GPRS was kept for EDGE Classic.

The basic GPRS radio concepts have not been modified. The logical channels that have been introduced for the GPRS system are reused for EGPRS. Data is still transferred over PDTCH, whereas signaling is transmitted over PACCH. The broadcast, control, and associated signaling channels are also exactly the same. The coding that is used over signaling channels is CS-1. This means that during a TBF an EGPRS mobile will transfer the block over its PACCH using CS-1 and the data will be transferred over PDTCH using MCS-1 to MCS-9.

The procedures that are used for power control and TA were retained. The MAC concept is also unchanged—mobiles can be multiplexed on the same physical channel. Note that EGPRS and GPRS mobiles can be multiplexed on the same PDCH. The concepts of TBF, TFI, and RR management are the same. The uplink multiplexing schemes such as dynamic allocation, extended dynamic allocation, and fixed allocation are unmodified. They can be used to multiplex GPRS and EGPRS mobiles on the same uplink PDCH. The signaling has been slightly changed to support dedicated EGPRS signaling during the establishment of a TBF.

The RLC Protocol is based on the same concept of sliding window. The same mechanism of segmentation has been kept, and the blocks are numbered with a BSN. Depending on the radio conditions, the link is adapted in such a way as to achieve the highest throughput.

## 2.3.2    New Modulation

The moderate throughput of high-speed circuit-switched data and GPRS is linked to the GMSK modulation and its limited spectrum efficiency. EDGE is based on a new modulation scheme that allows a much higher bit rate across the air interface. This modulation technique is called *eight-state phase-shift keying* (8-PSK). This modulation has an eight-state constellation allowing the coding of 3 bits per symbol (see Chapter 3 for more information). The raw bit rate is then three times higher than that for GMSK modulation.

The EGPRS transmitter adapts the modulation and CSs depending on the radio conditions; it can use GMSK or 8-PSK modulation according to the MCS used. The receiver is not informed of the modulation that is used by the transmitter. It has to perform blind detection of the modulation before being able to identify which MCS has been used, as described in Chapter 3.

Support of the 8-PSK modulation is mandatory for the mobile in downlink but is optional in uplink. On the network side, 8-PSK modulation is optional in both uplink and downlink. Thus, a network can support EDGE without implementing 8-PSK. In this case EDGE will not provide any gain in terms of maximum throughput but only some enhancements for the management of the radio link (RLC improvements). The significance of this solution is very limited as there is no gain in the maximum achievable throughput.

## 2.3.3    Link Quality Control

One of the main improvements of EGPRS, compared with GPRS, is in its link quality control. The enhancement was possible because of the introduction of a new ARQ scheme, *incremental redundancy* (IR), and new estimators for the link quality.

### 2.3.3.1    Coding Families

One imperfection of the GPRS system is linked to the design of the CSs. Before its transmission on the air interface, an LLC frame is segmented into data units with variable lengths. The length depends on the CS that will be used to transmit the radio block. Each CS was designed independently from the others and has its own data unit size. Once segmented, the data unit is sent over the air interface. If the receiver is not able to decode the radio block that carries this data unit, the transmitter will have to resend it later. The data unit can be retransmitted only with the same CS (i.e., at the same cod-

**Table 2.2**

MCS Families

| Family | Modulation and Coding Scheme |
|--------|------------------------------|
| A | MCS9 MCS6 MCS3 |
| A' | MCS8 MCS6 MCS3 |
| B | MCS7 MCS5 MCS2 |
| C | MCS4 MCS1 |

ing rate). If the radio conditions have changed or the coding rate is not appropriate to them, the receiver will never be able to decode the retransmission of the RLC data block. This will lead to the release of the TBF and the establishment of a new one in order to transmit the LLC frame.

In order to avoid this problem, the choice of the CS on the network side has to be made carefully. This often results in an unoptimized use of the radio interface, leading to a reduction of network capacity compared with its theoretical capacity.

The EGPRS modulation and CSs have been designed to offset this problem. The MCSs are divided into four families—A, A', B, and C. Each family includes several MCSs. A basic data unit with a fixed size is associated with each family. One radio block encoded with one MCS can carry one or several basic data units of its family. The sizes of the basic data unit for families A, A', B, and C are, respectively, 37, 34, 28, and 22 bytes. Table 2.2 lists the MCSs belonging to the various families.

MCS-1, MCS-2, and MCS-3 allow the carrying of one basic data unit of the respective families C, B, A', and A. MCS-4, MCS-5, and MCS-6 carry two basic data units, whereas MCS-7, MCS-8, and MCS-9 contain four basic data units. Different coding rates within a family are achieved by transmitting a different number of basic data units within one radio block.

Figure 2.2 shows the different families. For each of them, it gives the associated MCSs, the size of the basic data unit, and the number of data units that are carried by the different MCSs. The basic data unit size of family A' is 34 bytes. However, MCS-3 and MCS-6 are common to both families A and A'. That is why when a block is retransmitted from MCS-8 in MCS-6 or MCS-3, 3 padding bytes are added to the data part of the radio block.

One radio block sent by using MCS-9 is composed of four data units of 37 bytes. These data units can be retransmitted latter by the transmitter with two radio blocks by using MCS-6, leading to a two-times-lower coding

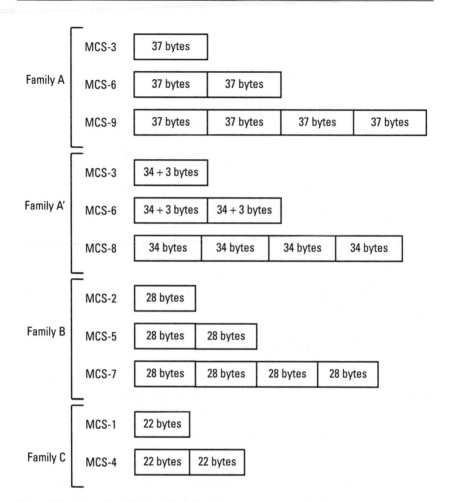

**Figure 2.2**   The various MCS families with their respective payload units.

rate compared with the previous transmission, or with four radio blocks by using MCS-3, leading to a four-times-lower coding rate. Every time one block is segmented and retransmitted with two separate radio blocks, the coding rate is decreased by two. That increases the capability to decode the block. MCS-7, MCS-8, and MCS-9 are composed of two RLC data blocks whereas the other MCSs contain only one RLC data block or half an RLC data block (this is described in Chapter 4).

This segmentation mechanism operates in connection with the link adaptation. When the network orders the use of a lower MCS compared with the initial one, the mobile will automatically segment the RLC data blocks that have to be retransmitted. The RLC data blocks that are sent for the first time will be transmitted with the last-ordered MCS.

### 2.3.3.2 IR

IR is an enhanced ARQ mechanism that reuses information from the previous transmissions of an RLC data block that was badly decoded in order to increase the capability to decode it when it is retransmitted. It consists of combining, at the output of the receiver demodulator, soft bits information from $N$ different transmissions of the same RLC blocks. This mechanism (described in greater detail in Chapter 4) can be associated with link adaptation in order to provide superior radio efficiency on the air interface.

### 2.3.3.3 Link Adaptation Measurements: BEP

The link adaptation mechanism relies on measurements performed by the mobile during a downlink transfer and by the BTS during an uplink transfer.

For GPRS the quality measurements that are used for link adaptation are based on RXQUAL. RXQUAL is limited to eight values, each corresponding to a range of BERs. The BER is calculated on a radio block basis and averaged. The returned RXQUAL corresponds to the range in which the averaged BER is mapped. The RXQUAL was originally designed for speech services. The different RXQUAL values were well suited to be mapped on the basis of speech quality, but this metric is not suitable for packet transfer. RXQUAL suffers from a lack of precision in the reported values. Moreover, because it is evaluated on a radio block basis (on four bursts), its meaning is not the same from one mobile environment to another.

During the standardization of EGPRS it was decided to introduce a new metric that is more suited to packet transfer by allowing a wider reporting range. This new metric is the *bit error probability* (BEP). Evaluated on a burst-by-burst basis, it gives more information about the mobile environment and its variations within a radio block. The range of the BEP is more significant and allows more precise reporting with a wider range.

## 2.3.4 RLC/MAC Improvements

In order to support increased throughput on the air interface, and the new coding types and to improve some GPRS weaknesses, the RLC/MAC protocol was modified. However the same basic concepts as for GPRS were retained.

The first improvement in the RLC protocol for EGPRS is the increase of the window size that is used for the control of the transmission of RLC data blocks. As described in Section 1.3.2.2 in Chapter 1, the RLC Protocol uses a fixed-size window of 64 for GPRS. This window size is well suited for packet transfer on one, two or three time slots (for more information see [1], Chapter 5). But it can be a limiting factor when a packet transfer occurs on additional time slots, leading to the stalling of the RLC Protocol. This issue is even more significant with the introduction of EGPRS, because one radio block encoded with MCS-7, MCS-8, or MCS-9 contains two RLC data blocks (see Chapter 4). This leads potentially to the transmission of twice as many RLC data blocks within the same period. To solve this problem, the RLC protocol was modified and variable-length windows were introduced. Depending on the number of allocated time slots, a more or less larger RLC window can be used. The range of this window is from 64 to 1,024 in steps of 64.

The increase of the window size had the result of increasing the reporting bitmap that is used by the receiver to acknowledge the decoded RLC data blocks. For GPRS, the window size is 64, and a reported bitmap of 64 can fit in one RLC/MAC control message encoded with CS-1. As described previously, EGPRS reuses the same coding as GPRS for the encoding of an RLC/MAC signaling message (CS-1). As a CS-1 radio block contains only 22 bytes of data information, it is not possible to report a bitmap that can contain up to 1,024 bits.

The second improvement concerns the reporting mechanism for EGPRS. It has been changed so that it can report the full RLC window status in several radio blocks. In order to provide more efficiency, the bitmap is also compressed to reduce its size. This reporting mechanism is described in Chapter 5.

These are the two major changes to the RLC Protocol. The MAC layer is also affected by some changes linked to the establishment of the TBF in EGPRS mode. This change is also described in Chapter 5.

## 2.3.5  RLC Data Block Format for EGPRS

The introduction of EGPRS called for many modifications to the radio interface. These changes necessitated a complete redefinition of the RLC data block that is used for EGPRS data transfer. For EGPRS, the RLC data block is composed of a RLC/MAC header and one or possibly two RLC data

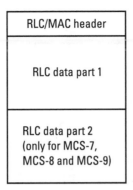

| RLC/MAC header |
| RLC data part 1 |
| RLC data part 2<br>(only for MCS-7,<br>MCS-8 and MCS-9) |

**Figure 2.3** RLC/MAC block structure for EGPRS data transfer.

parts, depending on the MCS that is used. MCS-7, MCS-8, and MCS-9 contain two RLC data parts, whereas the others contain only one. Figure 2.3 shows the RLC data block structure for EGPRS.

In EGPRS, the RLC/MAC header and the RLC data parts of the radio block do not have the same protection. To ensure strong header protection (because some information within the header is needed to decode the entire radio block and to perform IR), the header part of the radio block is independently coded from the data parts. When the radio block contains two RLC data parts, they are also independently coded, but at the same coding rate.

Three RLC/MAC header types have been defined. One is used to send two RLC data parts within one MCS-7, MCS-8, or MCS-9 radio block. The second is common to MCS-5 and MCS-6, and the last one to the GMSK MCSs. Note that for signaling, the same block format as for GPRS is used (CS-1).

### 2.3.5.1 RLC/MAC Block Header

*RLC/MAC Header for the Downlink*

*Common RLC/MAC Header for MCS-1, MCS-2, MCS-3, and MCS-4*  The common RLC/MAC header for the GMSK MCSs is given in Figure 2.4.

The RLC/MAC header contains the following fields.

- *USF:* It is used as an uplink multiplexing means when dynamic or extended dynamic allocation scheme is used.

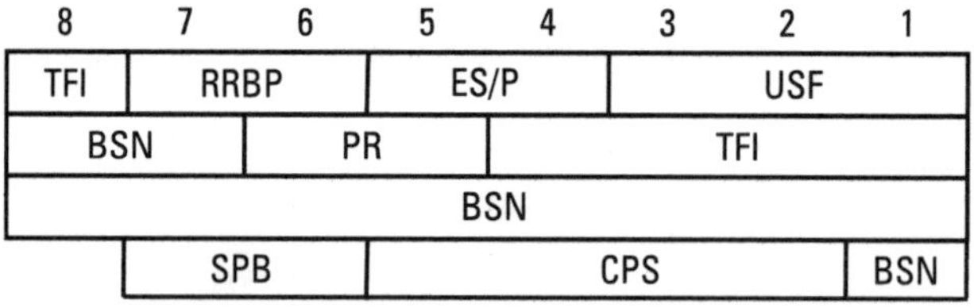

| 8 | 7 | 6 | 5 | 4 | 3 | 2 | 1 |
|---|---|---|---|---|---|---|---|
| TFI | RRBP | | ES/P | | USF | | |
| BSN | | PR | | TFI | | | |
| BSN | | | | | | | |
| | SPB | | CPS | | | | BSN |

**Figure 2.4**  Downlink RLC/MAC header for GMSK MCSs.

- *EGPRS supplementary/polling* (ES/P): It indicates whether the RRBP field is valid or not. If the RRBP is valid and if it is used to schedule the report of a downlink acknowledgment message, it indicates the types of parameters that are to be included in the message.
- *Relative Reserved Block Period* (RRBP): It indicates the number of frames that the mobile must wait for before transmitting an RLC/MAC control block.
- *TFI:* It identifies the downlink TBF.
- *Power reduction* (PR): It indicates the PR that was used by the BTS to transmit the current downlink block.
- *BSN:* It identifies the sequence number of the RLC block in the TBF.
- *Coding and puncturing scheme* (CPS): This field indicates the channel coding that was used (MCS-1, MCS-2, MCS-3, or MCS-4) and the puncturing scheme used.
- *Split block* (SPB): The SPB indicator is used when a RLC data block is segmented and retransmitted in two parts. It also indicates the first or second part of the segmented RLC data.

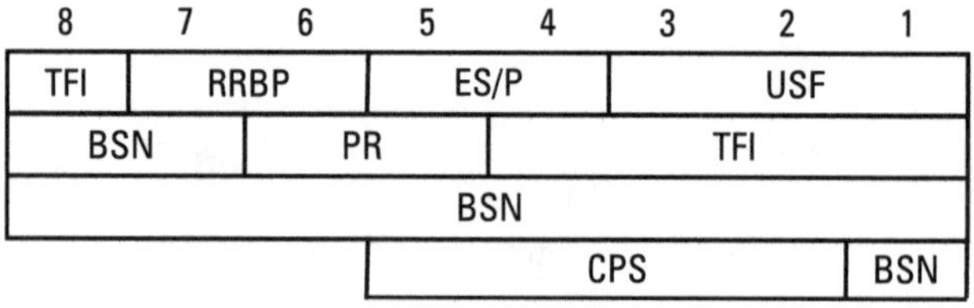

| 8 | 7 | 6 | 5 | 4 | 3 | 2 | 1 |
|---|---|---|---|---|---|---|---|
| TFI | RRBP | | ES/P | | USF | | |
| BSN | | PR | | TFI | | | |
| BSN | | | | | | | |
| | | | CPS | | | | BSN |

**Figure 2.5**  Downlink RLC/MAC header for MCS-5 and MCS-6.

| 8 | 7 | 6 | 5 | 4 | 3 | 2 | 1 |
|---|---|---|---|---|---|---|---|
| TFI | RRBP | | ES/P | | USF | | |
| BSN1 | | PR | | TFI | | | |
| BSN1 | | | | | | | |
| BSN2 | | | | | | | BSN1 |
| CPS | | | | BSN2 | | | |

**Figure 2.6**   Downlink RLC/MAC header for MCS-7, MCS-8, and MCS-9.

*Common RLC/MAC Header for MCS-5 and MCS-6*   The format of the RLC/MAC header for MCS-5 and MCS-6 is shown in Figure 2.5. It includes the same fields as in the previous section except for the SPB, which is not used with these MCSs. Note that the radio block contains only one RLC data part that is identified by the BSN field.

*Common RLC/MAC Header for MCS-7, MCS-8, and MCS-9*   The RLC/MAC header for MCS-7, MCS-8, and MCS-9 is shown in Figure 2.6. When a radio block is sent using these MCSs, it contains two RLC data parts that are identified by BSN1 and BSN2.

*RLC/MAC Header for the Uplink*

*Common RLC/MAC Header for MCS-1, MCS-2, MCS-3, and MCS-4*   The RLC/MAC header for the GMSK MCSs during uplink transfer is shown in Figure 2.7. The RLC/MAC header contains the following fields:

- *Retry* (R): It indicates whether the mobile transmits the access request message once or more than once during its most recent channel access.

| 8 | 7 | 6 | 5 | 4 | 3 | 2 | 1 |
|---|---|---|---|---|---|---|---|
| TFI | | Countdown value | | | | SI | R |
| BSN | | | | | | TFI | |
| CPS | | BSN | | | | | |
| | Spare | PI | RSB | SPB | | CPS | |

**Figure 2.7**   Uplink RLC/MAC header for GMSK MCSs.

- *Stall indicator* (SI): It indicates whether the MS RLC transmit window is stalled or not.

- *CV:* It indicates the number of RLC data block that remain to be sent before the end of the TBF.

- *TFI:* It identifies the uplink TBF.

- *BSN:* It indicates the sequence number of the RLC block in the TBF.

- *CPS:* This field indicates the channel coding and the puncturing that has been used for data.

- *SPB:* The SPB indicator is used when an RLC data block is segmented and retransmitted in two parts. It also indicates the first or second part of the segmented RLC data.

- *Resent block bit* (RSB): This bit is used to indicate whether the RLC data part is being sent for the first time or not.

- *PFI indicator* (PI): It indicates the presence of the optional byte that contains the PFI within the data part.

*Common RLC/MAC Header for MCS-5 and MCS-6*　　MCS-5 and MCS-6 RLC/MAC headers are shown in Figure 2.8.

*Common RLC/MAC Header for MCS-7, MCS-8, and MCS-9*　　Figure 2.9 shows the uplink RLC/MAC header for MCS-7, MCS-8, and MCS-9. Note that two RLC data parts are contained within one MCS-7, MCS-8, or MCS-9 radio blocks. They are identified by BSN1 and BSN2.

| 8 | 7 | 6 | 5 | 4 | 3 | 2 | 1 |
|---|---|---|---|---|---|---|---|
| TFI | | Countdown value | | | | SI | R |
| BSN | | | | | TFI | | |
| CPS | | BSN | | | | | |
| Spare | | | | | PI | RSB | CPS |
| | | | Spare | | | | |

**Figure 2.8**　Uplink RLC/MAC header for MCS-5 and MCS-6.

| 8 | 7 | 6 | 5 | 4 | 3 | 2 | 1 |
|---|---|---|---|---|---|---|---|
| TFI | | Countdown value | | | | SI | R |
| BSN1 | | | | | TFI | | |
| BSN2 | | BSN1 | | | | | |
| BSN2 | | | | | | | |
| Spare | PI | RSB | | CPS | | | |
| Spare | | | | | | | |

**Figure 2.9**   Uplink RLC/MAC header for MCS-7, MCS-8, and MCS-9.

### 2.3.5.2   RLC Data Part

The EGPRS RLC data part is shown in Figure 2.10. It is composed of two bits and the RLC data unit. The *extension* (E) bit is used to indicate the presence of an optional byte in the RLC data unit. The *final block indicator* (FBI) bit is used only in the downlink direction to indicate that the block is the final one of the TBF. The *TLLI indicator* (TI) bit is used only in the uplink direction by the mobile to indicate the presence of the TLLI field within the RLC data unit. The *length indicator* (LI) field is used to delimit LLC PDUs within the RLC data unit.

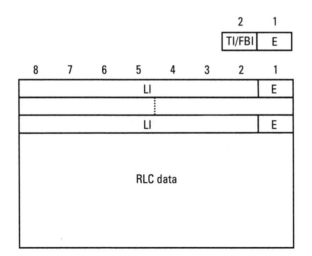

**Figure 2.10**   EGPRS RLC data part.

**Table 2.3**
EGPRS RLC Data Part Size

| MCS | EGPRS RLC Data Unit Size in Bytes | Additional Bit in the Data Part |
|-----|-----------------------------------|--------------------------------|
| MCS-1 | 22 | 2 |
| MCS-2 | 28 | 2 |
| MCS-3 | 37 | 2 |
| MCS-4 | 44 | 2 |
| MCS-5 | 56 | 2 |
| MCS-6 | 74 | 2 |
| MCS-7 | 2*56 | 2*2 |
| MCS-8 | 2*68 | 2*2 |
| MCS-9 | 2*74 | 2*2 |

For MCS-7, MCS-8, and MCS-9, the radio block contains two EGPRS RLC data parts. Table 2.3 gives the size of the EGPRS data unit depending on the MCS.

# Reference

[1]   Seurre, E., P. Savelli, and P. J. Pietri, *GPRS for Mobile Internet*, Norwood, MA: Artech House, 2003.

# Selected Bibliography

3GPP TS 03.64 Overall Description of the GPRS Radio Interface, Stage 2 (R99).

3GPP TS 04.18 Radio Resource Control Protocol (R99).

3GPP TS 04.60 Radio Link Control/Medium Access Control (RLC/MAC) Protocol (R99).

3GPP TS 05.01 Physical Layer on the Radio Path; General Description (R99).

3GPP TS 05.02 Multiplexing and Multiple Access on the Radio Path (R99).

# 3

# RF Physical Layer

As seen in Chapter 2, the general principles of the GPRS radio interface have been kept in EGPRS. The physical channel definition, the multiframe organization, and the logical channels and their mapping onto physical channels are unchanged.

The major modification is the introduction of a new modulation scheme in addition to GMSK: It is the EDGE 8-PSK modulation, the aim of which is to increase the achievable data bit rate. This chapter is dedicated to this major change, with a presentation of its impacts on system requirements. It is illustrated with case studies to make the reader aware of the implementation constraints concerning the RF transceiver, as well as the digital generation of the modulation signal in the transmitter and the digital demodulator in the receiver.

First, the GMSK and 8-PSK modulations are presented in detail. This presentation provides the opportunity to discuss the accuracy requirements on the modulated signal at the output of the transmitter. Next, we present the modifications that this modulation imposes on the RF transmitter specifications: the 8-PSK power classes, spectrum due to the modulation, and power template (variations of the RF power during the transmitted burst). On the receiver side, the sensitivity and interference performance requirements of EGPRS are briefly introduced, as well as the *nominal error rate* (NER). The end of the chapter presents case studies concerning the modulation signal generation, the general RF constraints, and the demodulation aspects.

## 3.1   Modulation

### 3.1.1   GMSK Modulation Overview

Digital modulation consists of the variation of either the amplitude, the phase, or the frequency of a carrier or a combination of these three characteristics. The *Gaussian minimum shift keying* (GMSK) belongs to a subset of phase modulations, namely, the *continuous phase modulation* (CPM) family.

In this family, the modulated signal $m(t)$ can be expressed as [1]:

$$m(t) = A_0 \cos\left(2\pi f_0 t + \varphi(t)\right) \tag{3.1}$$

where $A_0$ is the amplitude, which does not vary with the input data bits; $f_0$ is the carrier frequency; and $\varphi(t)$ is the phase of the signal that carries the modulation information.

At a given instant, the phase $\varphi(t)$ is equal to

$$\varphi(t) = 2\pi h \sum_{k=-\infty}^{+\infty} \alpha_k q(t - kT) \tag{3.2}$$

where $\alpha_k = \pm 1$ is the $k$th transmitted symbol, $T$ is the symbol period, $h$ is the so-called modulation index, and $q(t)$ is the integral of a pulse, $g(t)$:

$$q(t) = \int_0^t g(\tau) d\tau \tag{3.3}$$

such that

$$\lim_{t \to +\infty} q(t) = 1/2$$

An example of pulse is given in Figure 3.1. In the left part of the figure is represented a rectangle $rect_T$ of amplitude $1/2T$ and duration $T$. The $g(t)$ is equal to

$$\frac{1}{2T} rect_T \left( t - T/2 \right)$$

The right part of the figure shows $q(t)$, which is obtained by integration of $g(t)$.

The phase modulated signal is therefore

$$m(t) = A_0 \cos\big(\phi(t)\big) = A_0 \cos\left( 2\pi f_0 t + 2\pi h \sum_{k=-\infty}^{+\infty} \alpha_k q(t - kT) \right) \quad (3.4)$$

with

$$\phi(t) = 2\pi f_0 t + 2\pi h \sum_{k=-\infty}^{+\infty} \alpha_k q(t - kT)$$

Since the phase is continuous, it can also be seen as a frequency-modulated signal, and the instantaneous frequency is

$$f_i(t) = \frac{1}{2\pi} \frac{d\phi}{dt} = f_0 + h \sum_{k=-\infty}^{+\infty} \alpha_k g(t - kT) \quad (3.5)$$

For example, the *continuous phase frequency shift keying* (CPFSK) modulation is based on the pulse $g(t)$ depicted in Figure 3.1, which is a rectangle of duration $T$:

$$g(t) = \frac{1}{2T} rect_T \left( t - \frac{T}{2} \right) \quad (3.6)$$

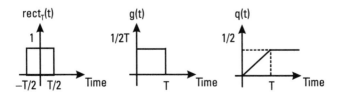

**Figure 3.1**   Example of a pulse in CPM modulations.

The elementary phase increment between two symbols,

$$2\pi h \int_0^t g(\tau)d\tau$$

is:

$$\begin{cases} \dfrac{\pi h}{T} t \text{ for } 0 \leq t \leq T \\ \pi h \text{ for } t > T \\ 0 \text{ for } t < 0 \end{cases} \qquad (3.7)$$

The *minimum shift keying* (MSK) is a CPFSK, where the modulation index $h$ is equal to ½. This means that between two input symbols, the phase variation is $\pm \pi/2$, as shown in Figure 3.2. In this example, the bold line shows a phase trajectory for the input sequence $(\alpha_k) = (1, 1, 1, -1, 1, 1)$, taking into account that both phases $+\pi$ and $-\pi$ represent actually the same state on the trajectory.

The GMSK modulation used in the GSM system is a modified MSK, in which the impulse $g(t)$ is a convolution of a rectangle with a Gaussian filter $h(t)$:

$$\begin{cases} g(t) = rect_T\left(t - \dfrac{T}{2}\right) * h(t) \\ h(t) = \sqrt{\dfrac{2\pi}{\ln(2)}} B \exp\left(-\dfrac{2\pi^2 B^2}{\ln(2)} t^2\right) \\ \varphi(t) = 2\pi h \int_{-\infty}^{t} \sum_{k=-\infty}^{+\infty} \alpha_k g(\tau - kT) d\tau = \pi \sum_{k=-\infty}^{+\infty} \alpha_k q(\tau - kT) \end{cases} \qquad (3.8)$$

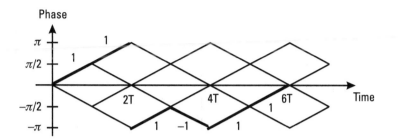

**Figure 3.2** MSK modulated signal phase variations during time.

In (3.8) $B$ represents the 3-dB bandwidth of the Gaussian filter. Indeed, in the frequency domain, the Gaussian filter $h(t)$ can be defined by:

$$H(f) = \exp\left(-\frac{\ln(2)f^2}{2B^2}\right) \tag{3.9}$$

and $g(t)$ may also be represented as follows:

$$g(t) = \frac{1}{2T}\left(Q\left(2\pi B \frac{t - T/2}{\sqrt{\ln(2)}}\right) - Q\left(2\pi B \frac{t + T/2}{\sqrt{\ln(2)}}\right)\right) \tag{3.10}$$

with $Q(t)$ defined in (3.16).

In the GSM system, $B$ has been chosen such that the product $BT$, also called the normalized bandwidth, is equal to 0.3.

This choice allows a good trade-off between the spectral efficiency of the modulation and the BER performance that can be achieved in the receiver. Indeed, the Gaussian filtering reduces the modulation bandwidth, but it may introduce a degradation in terms of receiver performance.

Figure 3.3 is a graphical representation of the impulse response $g(t)$, for different values of the product $BT$. Note that for an infinite value of $BT$, $g(t)$ corresponds to the MSK modulation pulse. This is because for a given value of $T$, $B$ tends to infinity, and $h(t)$ to the Dirac pulse $\delta(t)$. Therefore, the pulse $g(t)$ tends to a rectangle, which is the case with MSK.

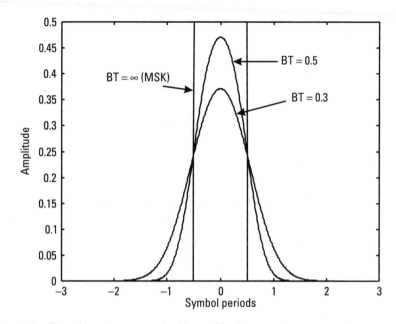

**Figure 3.3**    The $g(t)$ impulse response for different $BT$ values.

The advantage of GMSK compared to MSK is the spectral efficiency, as mentioned previously. Indeed, the use of the Gaussian filter makes is possible to reduce the level of the side lobes, as can be seen in Figure 3.4. In a real implementation of the GMSK modulator and RF transmitter, many imperfections affect the quality of the signal. For instance, some quantification noise due to the digital-to-analog converters or some phase noise from the synthesizer can affect the GMSK signal.

The quality of the modulated signal is measured by its phase error. The phase error of the modulation is determined by comparing the transmitted RF modulated signal with an ideal GMSK signal. The different steps of the estimation are as follows (see Figure 3.5):

- First, the output of the transmitter is downconverted to the baseband I and Q signals.
- The phase of this signal is estimated (it is the argument of the complex signal $I + j \cdot Q$). The frequency error of the transmitter, if there is any, is suppressed to this phase signal. Mathematically, correcting

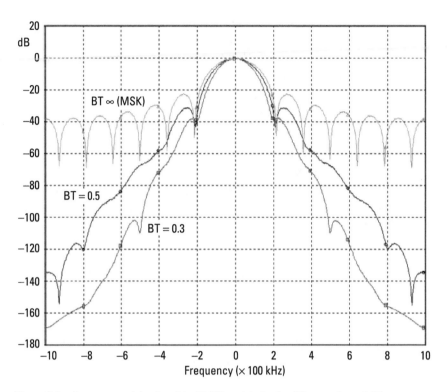

**Figure 3.4** Power spectral density of the GMSK modulation for different values of *BT*.

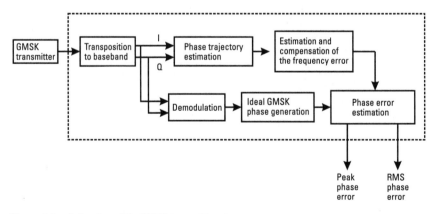

**Figure 3.5** Estimation of the GMSK transmitter phase error.

the frequency error means removing a linear component of the phase signal. This suppression is due to the fact that the frequency error tolerance is specified independently from the phase error.

- In parallel, this downconverted signal is demodulated to retrieve the input bit sequence. This binary sequence is used to regenerate an ideal GMSK modulated signal, which is equal to the mathematical expression of the modulation signal.

- The difference between the received phase signal and the ideal GMSK phase signal is calculated.

- These phase difference samples represent the instantaneous phase-error signal (the difference between the phase of the transmitted waveform and the phase of the expected one), which is used to estimate the peak and *root mean square* (rms) phase errors.

The specifications state that the rms phase error shall not be greater than 5°, and the maximum peak deviation during the burst shall be less than 20°.

### 3.1.2   8-PSK Modulation

#### 3.1.2.1   Description of the Modulation

In the EDGE system, in addition to the GMSK modulation, a modified 8-PSK scheme can also be employed. *M-state phase modulations* (M-PSK modulations) are digital modulations in which the information is conveyed on the phase of the carrier. For M-PSK, the modulated signal $m(t)$ may be expressed as

$$m(t) = A_0 \cos\left(2\pi f_0 t + \varphi(t)\right) \tag{3.11}$$

In (3.11) $A_0$ and $f_0$ are the amplitude and frequency of the carrier, and

$$\varphi(t) = \sum_k \phi_k \delta(t - kT)$$

represents the phase modulation signal. $T$ is the symbol duration, and $\phi_k$ is the $k$th modulating symbol, which can take $M$ different values: $\phi_k = \theta_0 + 2.m.\pi/M$, where $m \in [0, M-1]$, and $\theta_0$ is an offset.

The $\delta(t)$ is the Dirac pulse:

$$\begin{cases} \delta(0) = 1 \\ \delta(t) = 0 \text{ for } t \neq 0 \end{cases}$$

In this scheme, each symbol $\phi_k$ carries $n$ information bits, where $n = \log_2 M$.

Hence, we have

$$\begin{aligned} m(t) &= A_0 \cos\left( 2\pi f_0 t + \sum_k \phi_k \delta(t - kT) \right) \\ &= A_0 \sum_k \left[ \cos(\phi_k).\cos(2\pi f_0 t) - \sin(\phi_k).\sin(2\pi f_0 t) \right] \\ &\quad .\delta(t - kT) \\ &= \sum_k \left[ I_k.\cos(2\pi f_0 t) - Q_k.\sin(2\pi f_0 t) \right].\delta(t - kT) \end{aligned} \tag{3.12}$$

where $I_k = \cos(\phi_k)$ and $Q_k = \sin(\phi_k)$ are the in-phase and in-quadrature signals, respectively.

This signal is then passed through a shaping filter, which may, for instance, be the rectangle of duration $T$:

$$m(t) * rect_T(t) = \sum_k \left[ I_k.\cos(2\pi f_0 t) - Q_k.\sin(2\pi f_0 t) \right].rect_T(t - kT)$$

Note that the EDGE pulse-shaping filter is different from this example, as we will see later on in this description.

Figure 3.6 represents the PSK modulation symbols in the complex plane for $M = 2$, 4, and 8. For 8-PSK modulation, each symbol corresponds to three consecutive bits. The modulated signal phase may take 8 values, $2.m.\pi/8$, for $0 \leq m \leq 7$. Each value is associated with a symbol, representing 3 bits.

The correspondence between the symbols and the 3-bit sequences in the case of EDGE is shown in Figure 3.7. In this figure, each symbol differs from the two nearest symbols by only one bit.

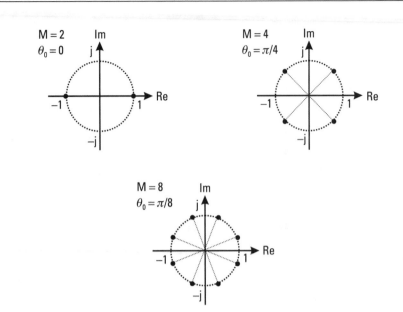

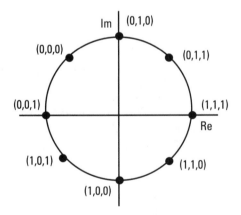

**Figure 3.6**  Example of M-PSK modulations, for $M$ = 2, 4, and 8.

This method, called the Gray mapping, minimizes the BER in the case where a received symbol is misinterpreted due to noise or interference (most of the time, when the receiver makes an error, it replaces the transmitted symbol by one of its neighbors).

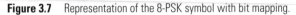

**Figure 3.7**  Representation of the 8-PSK symbol with bit mapping.

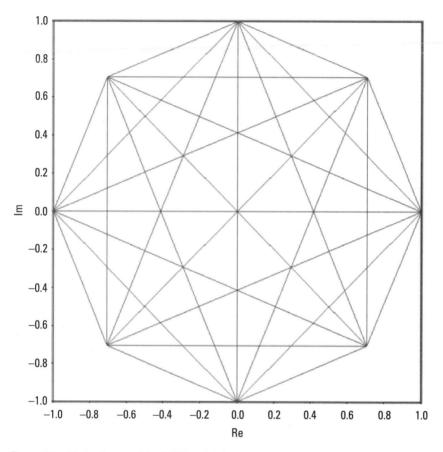

**Figure 3.8** Vector diagram of the 8-PSK modulation.

In Figure 3.8 a vector diagram is shown, which is a representation of all the possible transitions between symbols. For EDGE, it was decided to use a modified 8-PSK scheme, in which the vector diagram trajectory does not cross the origin (I and Q signals equal to zero). This is explained by implementation constraints concerning the power amplifiers, as discussed in the case study of Section 3.4.3. Indeed, zero crossings in the constellation increase the envelope power variations, which increases the distortion when the signal is passed through a power amplifier.

In order to avoid zero crossings due to the π radians phase shifts, the solution that was chosen consists of rotating the constellation of 3π/8 at each

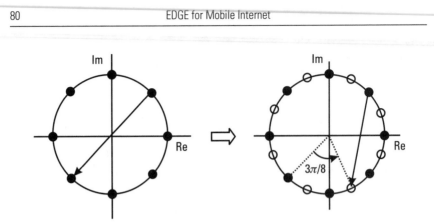

**Figure 3.9**    The $3\pi/8$ phase rotation to avoid zero crossings.

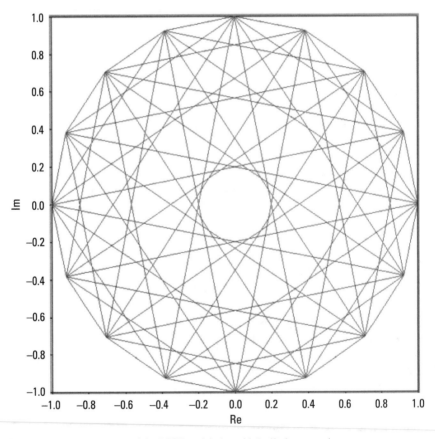

**Figure 3.10**  Vector diagram of the 8-PSK modulation with $3\pi/8$ phase rotation.

symbol period (see Figure 3.9). Figure 3.10 displays the vector diagram of the modulation with this continuous offset of $3\pi/8$ radians. We see that this procedure generates eight new symbols in the constellation. The phase rotation results in a maximum phase shift $7\pi/8$ between two consecutive symbols, instead of $\pi$. This scheme is often referred to as offset 8-PSK modulation.

In order to attenuate the level of side lobes, the signal is filtered. The filter that was chosen is the so-called linearized GMSK pulse (i.e., the main component in the Laurent decomposition of the GMSK modulation). This means that the GMSK modulation can be approximated as a linear modulation in which the input symbols are rotated and then filtered by a pulse, instead of a nonlinear modulation, where the phase signal is filtered. This equivalent pulse has been calculated by Laurent (see [2] for further details on this topic).

The impulse response of this filter is defined by:

$$
c_0(t) = \begin{cases} \displaystyle\prod_{i=0}^{3} S(t+iT), \text{for } 0 \le t \le 5T \\[2mm] 0, \text{else} \end{cases} \tag{3.13}
$$

with

$$
S(t) = \begin{cases} \sin\left(\pi \displaystyle\int_0^t g(t')\,dt'\right), \text{for } 0 < t \le 4T \\[3mm] \sin\left(\dfrac{\pi}{2} - \pi \displaystyle\int_0^{t-4T} g(t')\,dt'\right), \text{for } 4T < t \le 8T \\[3mm] 0, \text{else} \end{cases} \tag{3.14}
$$

$$
g(t) = \frac{1}{2T}\left( Q\left(2\pi \cdot 0.3 \frac{t - 5T/2}{T\sqrt{\ln(2)}}\right) - Q\left(2\pi \cdot 0.3 \frac{t - 3T/2}{T\sqrt{\ln(2)}}\right)\right) \tag{3.15}
$$

and

$$Q(t) = \frac{1}{\sqrt{2\pi}} \int_t^{\infty} e^{-\frac{\tau^2}{2}} d\tau \tag{3.16}$$

Graphically, this filter impulse response is represented in Figure 3.11, and its effects on the spectrum occupancy of the modulation and on the vector diagram are shown in Figures 3.12 and 3.13.

One important point is that the filtering of the 8-PSK signal introduces some variation in its instantaneous power. Indeed, in the nonfiltered 8-PSK scheme, the phase changes are instantaneous between symbols at every $T$, which is not the case when some filtering is added. Figure 3.14 presents the power variations of the filtered signal, to show that the modulated signal envelope is not constant.

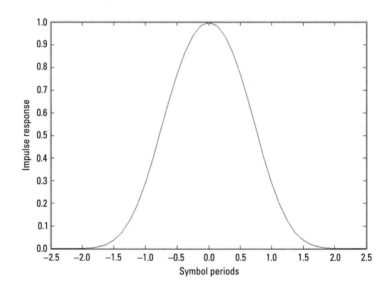

**Figure 3.11**  EDGE pulse-shaping filter impulse response.

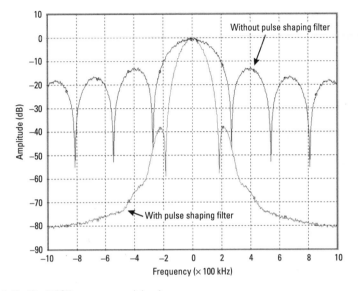

**Figure 3.12** The 8-PSK power spectral density.

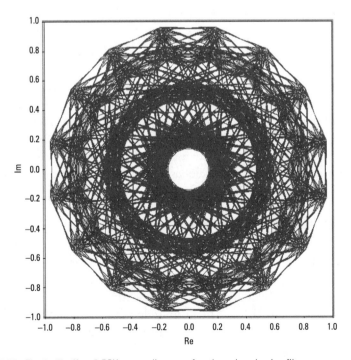

**Figure 3.13** The 3π/8-offset 8-PSK vector diagram after the pulse-shaping filter.

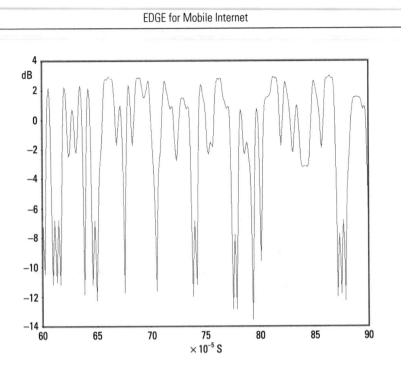

**Figure 3.14** Envelope power variation of the EDGE 8-PSK modulation.

**Note** The distance between two symbols in the 8-PSK constellation is smaller than in the GMSK one for a given energy per symbol. This characteristic increases the probability of misinterpretation of the symbols in the receiver, owing to the noise and interference. If the radio conditions are good there is no problem, and a greater data rate can be achieved by using 8-PSK. In the case of bad conditions, the performances are degraded with 8-PSK, and the use of GMSK may be required. This is the reason why the two modulations are used in EGPRS, and the adaptation of the modulation to the propagation conditions is based on measurements performed by the MS and BTS and controlled by the network (see Chapter 4).

### 3.1.2.2  The 8-PSK Modulation Accuracy

For the EDGE-8-PSK modulation, due to the nonconstant envelope nature of the modulated signal, the accuracy is defined by the error vector $E(k)$, which measures the difference between the vector representing the actual transmitted signal $Z(k)$ and the vector representing the error-free modulated signal $S(k)$ (see Figure 3.15). The magnitude of the error vector is called the *error vector magnitude* (EVM). The EVM specification is given in percentage relative to the magnitude of the error-free vector.

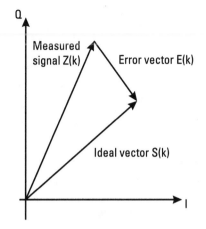

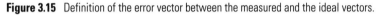

**Figure 3.15** Definition of the error vector between the measured and the ideal vectors.

The EVM is computed for each symbol period, and the following performance figures are derived:

- The rms EVM over the burst.

- The peak value of EVM is the peak error deviation within a burst, measured at each symbol interval.

- The ninety fifth percentile is defined as an EVM percentage below which 95% of the individual EVM values should be measured at each symbol interval. This means that no more than 5% of the symbols may have an EVM exceeding the ninety-fifth percentile point. Note that the EVM values are acquired during the burst, tail bits not included.

- The *origin offset suppression* (OOS) is a measure of the carrier feedthrough, which is the relative power level of the RF modulated signal to the residual RF carrier sine wave signal. This residual signal may be due to a leakage from the RF transmitter in some architectures (see [3], Section 4.3.4).

The maximums allowed for these values are given in Table 3.1, for normal (+15°C to +35°C) and extreme conditions (–10°C to +55°C).

The principle of the EVM (peak, rms, and ninety-fifth percentile) and OOS measurements is shown in Figure 3.16.

**Table 3.1**
EVM Specifications

| For the MS | | |
| --- | --- | --- |
| | **Normal Conditions** | **Extreme Conditions** |
| rms EVM | ≤ 9% | ≤ 10% |
| Peak EVM | ≤ 30% | ≤ 30% |
| 95th percentile EVM | ≤ 15 | ≤ 15% |
| Origin Offset Suppression (OOS) | ≥ 30 dB | ≥ 30 dB |
| **For the BTS** | | |
| | **Normal Conditions** | **Extreme Conditions** |
| rms EVM | ≤ 7% | ≤ 8% |
| Peak EVM | ≤ 22% | ≤ 22% |
| 95th percentile EVM | ≤ 11% | ≤ 11% |
| Origin Offset Suppression (OOS) | ≥ 35 dB | ≥ 35 dB |

- The estimator input consists of the digitized I and Q samples $Z(k)$ from the measured signal and $S(k)$ from the ideal EDGE modulated signal. Prior to this estimation, the amplitude of the signal $Z(k)$ is normalized to be the same as the ideal reference signal. In addition, a constant phase rotation is applied to the samples $Z(k)$, if needed, to compensate for a possible rotation due to the transmitter *local oscillator* (LO) initial phase. This operation of normalization and phase rotation is symbolized by the division by the complex factor $C_1$ in Figure 3.16. Also, the frequency error of the transmitter, as well as any amplitude drift, is corrected at this stage (complex factor $W^{-t}$).

- After these compensations, the signal is passed through a measurement filter and sampled at the symbol rate (270.833 kHz) to produce the signal $Z(k) = S(k) + E(k) + C_0$, where $C_0$ is a constant origin offset, representing the carrier feedthrough.

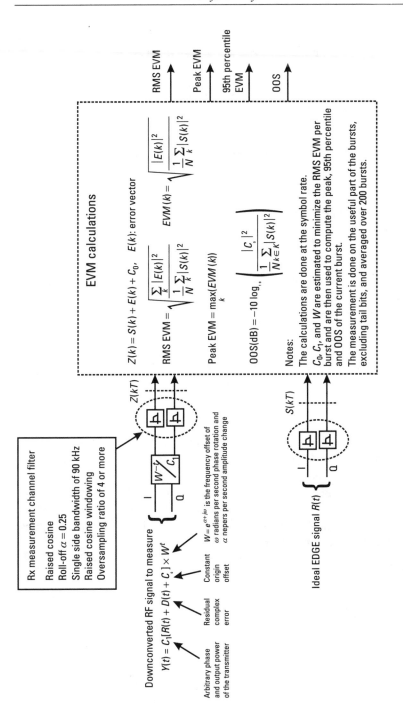

**Figure 3.16** Procedure for the EVM estimation.

- The error vector $E(k) = Z(k) - C_0 - S(k)$ is measured and calculated for each instant $k$. The rms EVM is defined as:

$$\text{rms EVM} = \sqrt{\sum_k |E(k)|^2 \Big/ \sum_k |S(k)|^2} \tag{3.17}$$

The symbol EVM at symbol $k$ is defined as

$$\text{EVM}(k) = \sqrt{\frac{|E(k)|^2}{\frac{1}{N}\sum_k |S(k)|^2}} \tag{3.18}$$

where $N$ is the number of samples.

- The OOS (in decibels) is defined as

$$\text{OOS (dB)} = -10\log_{10}\left(\frac{|C_0|^2}{\frac{1}{N}\sum_k |S(k)|^2}\right) \tag{3.19}$$

**Note**  $C_0$, $C_1$, and $W$ shall be estimated to minimize rms EVM per burst and then used to compute the individual vector errors $E(k)$ on each symbol.

In the above description, the errors shall be measured after a measurement filter that is defined by the specifications. It is a raised-cosine filter with roll-off 0.25 and single side-band bandwidth of 90 kHz. Its impulse response is multiplied by a window that is defined as:

$$w(t) = \begin{cases} 1, & 0 \le |t| \le 1.5T \\ 0.5\left(1+\cos\left[\pi\left(|t|-1.5T\right)\big/2.25T\right]\right), & 1.5T \le |t| \le 3.75T \\ 0, & |t| \ge 3.75T \end{cases} \tag{3.20}$$

## 3.2 RF Characteristics on the Transmitter Side

### 3.2.1 MS Power Classes

We have seen in [3] (Section 1.5.6.1) that several classes of mobiles exist in GSM, according to their maximum output power capability. EDGE mobiles that support the 8-PSK modulation on the transmit path may also belong to different 8-PSK power classes—E1, E2, and E3—presented in Table 3.2.

Of course, uplink power control is also present in 8-PSKs to reduce the MS transmitted power level according to commands that are sent by the network.

The GMSK and 8-PSK maximum output power for a given mobile are not necessarily the same, but the maximum output power for 8-PSK in any frequency band is always equal to or less than the GMSK maximum output power in the same band.

Typical handsets will probably be in the E2 power class, which corresponds to a lower maximum output power than the usual GSM MS power class. Indeed, most of the GSM900 mobiles available on the market are class 4 terminals (maximum output power: +33 dBm), and most of the DCS1800 mobiles are part of class 1 (maximum output power: +30 dBm). This reduction of the maximum output power is due to higher implementation constraints concerning the 8-PSK modulation, as explained in the case study given in Section 3.4.3.

For the BTS *transceiver* (TRX), the power classes given in [3], Chapter 1, for GMSK also exist in 8-PSK (there are no other power classes introduced for EDGE), but the manufacturer may declare a different power class for each modulation.

**Table 3.2**
8-PSK MS Power Classes

| Power Class | GSM400, GSM850, GSM900 Nominal Maximum Output Power | DCS1800 Nominal Maximum Output Power | PCS1900 Nominal Maximum Output Power |
|---|---|---|---|
| E1 | 33 dBm | 30 dBm | 30 dBm |
| E2 | 27 dBm | 26 dBm | 26 dBm |
| E3 | 23 dBm | 22 dBm | 22 dBm |

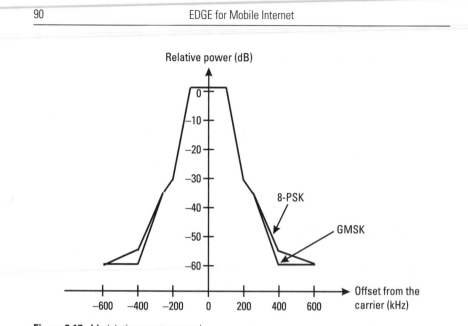

**Figure 3.17** Modulation spectrum mask.

## 3.2.2 Spectrum Due to Modulation

This requirement gives the spectrum mask that needs to be satisfied by the modulated signal. Compared with the GMSK spectrum mask, there is a 6-dB relaxation at 400 kHz from the carrier (see Figure 3.17). This is due to the spectral occupancy of the EDGE 8-PSK modulation.

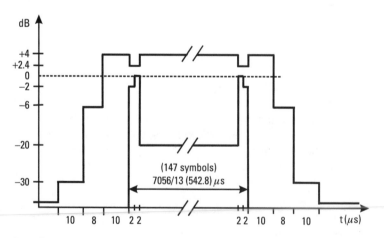

**Figure 3.18** Time mask for normal bursts with 8-PSK modulation. (*From:* [4]. © ETSI 2001.)

### 3.2.3 Power Versus Time Requirement

The transmitted power level relative to time when sending an EDGE 8-PSK burst is shown in Figure 3.18. The maximum power variation tolerance during the burst is +4/–20 dB relative to the average power (the 0-dB reference level corresponds to the average output power level). This is, of course, due to the nonconstant envelope characteristic of the 8-PSK, as discussed in Section 3.1.2.1.

## 3.3 RF Characteristics on the Receiver Side

### 3.3.1 EGPRS Sensitivity and Interference Performance

As with GPRS, the performance criterion to be met in EGPRS is the *block error rate* (BLER). It refers to all erroneously decoded data blocks including any wrong detection of the modulation and incorrect decoding of the stealing flags, header, data, or parity bits.

For PDTCH the BLER refers to RLC blocks. Therefore, for the EGPRS MCS7, MCS8, and MCS9 radio blocks, which carry two RLC data blocks (see Chapter 4), there can be up to two block errors per 20 ms radio block.

The reference BLER level for PDCH is 10%, except for some few cases where the requirement is 30%.

The sensitivity and interference performances are specified in a similar way to GPRS. The performance tables are not presented here for simplicity reasons, but all the detailed data can be found in [4].

### 3.3.2 8-PSK NER

The sensitivity performance specifies the minimum level of signal for which the 10% BLER is to be met. However, it does not specify up to which level of signal the receiver shall be able to demodulate.

This is the purpose of the NER specification. It correspond to a raw BER (i.e., the BER at the equalizer output) limit, in static channel conditions. The specification is as follows:

- The raw BER shall be lower than $10^{-4}$ for signal levels higher than –84 dBm for the BTS, and –82 dBm for the MS, and up to –40 dBm.

**Table 3.3**
8-PSK BER Performance with a Frequency Error

| | Frequency Offset | BER | Input Signal Level |
|---|---|---|---|
| GSM400 normal BTS | 0.2 ppm | $\leq 10^{-4}$ | $\geq -84$ dBm |
| GSM850 and GSM900 normal BTS | 0.1 ppm | $\leq 10^{-4}$ | $\geq -84$ dBm |
| DCS1800, PCS1900 normal BTS | 0.1 ppm | $\leq 10^{-4}$ | $\geq -84$ dBm |
| GSM400, GSM850, and GSM900 MS | 0.1 ppm | $\leq 10^{-4}$ | $\geq -82$ dBm |
| DCS1800 and PCS1900 MS | 0.1 ppm | $\leq 10^{-3}$ | $\geq -82$ dBm |

- A BER of $10^{-3}$ is to be maintained for both MS and BTS, for input levels up to $-26$ dBm (the BER must remain acceptable for high receive signal levels, near receiver saturation conditions).

- When the frequency of the 8-PSK modulated signal is offset by a fixed frequency error, with a random sign at each received burst, the performance given in Table 3.3 shall be met. In this table the frequency offset is given, with the BER requirement and the *input signal level* (ISL). All these requirements are valid up to $-40$ dBm.

### 3.3.3  Modulation Detection

It is to be noted that in EGPRS, the modulation that is used for a particular block, in either uplink or downlink, is not necessarily known in advance by the receiver. Indeed, to allow for fast switching between GMSK and 8-PSK modulations—and thus a good reaction time in varying radio conditions—it was decided to avoid the notification of the modulation change in upper layer messages. This also avoids any signaling overhead. Instead, a mechanism has to be implemented in the receiver to determine whether GMSK or 8-PSK was used, before the equalization can take place. This mechanism is called blind detection of the modulation. This section describes how blind detection is performed.

#### 3.3.3.1  Format of the 8-PSK Normal Burst (NB)

The NB is defined for both GMSK and 8-PSK modulation schemes. The 8-PSK NB, shown in Figure 3.19, is very similar to the GMSK NB presented in ([3], Chapter 1), except that a symbol period carries 3 bits instead of 1 for GMSK.

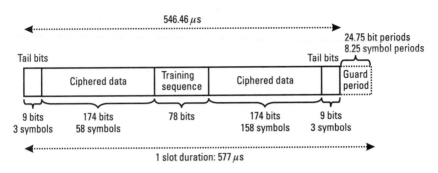

**Figure 3.19** Structure of the 8-PSK NB.

### 3.3.3.2  8-PSK Training Sequence Codes (TSCs)

For the 8-PSK modulation, there are eight training sequences, numbered from 0 to 7, as for GMSK. It is important to notice that the bit sequence for a given TSC in 8-PSK can be obtained from the corresponding GMSK TSC bits by replacing a 1 with 001, and a 0 with 111. In Figure 3.7, the bit sequence 001 corresponds to the symbol −1 on the 8-PSK constellation, and 111 corresponds to +1. This means that symbol wise, the training sequences are the same for 8-PSK as for GMSK; although 8-PSK modulation is complex, only the two real symbols +1 and −1 are used. This characteristic is intended to allow the blind detection of the modulation.

Indeed, in the EDGE system, the training sequences are not only used for channel estimation and time synchronization purposes, but also to determine which of the GMSK or 8-PSK is being used on the radio link.

### 3.3.3.3  Blind Detection of the Modulation

In the 8-PSK NB, the bit sequences (001) and (111) used for the training sequence are input to the EDGE 8-PSK modulator, with the following steps:

- The mapping of the bits onto the symbols of the constellation produces one of the two symbols, +1 and −1.
- A rotation of $3\pi/8$ radians per symbol is applied on the resulting signal.
- The symbols are filtered by the Gaussian pulse.

The resulting signal is very similar to the GMSK in the sense that each +1 symbol results in a phase shift of $3\pi/8$, and each −1 symbol results in a $-3\pi/8$ phase shift, whereas with GMSK a $\pm 1$ results in a $\pm \pi/2$ phase rotation.

This difference between the two modulation signals is used in the channel estimation stage of the receiver (see Section 3.4.4) to perform the blind detection of the modulation. For instance, the receiver can make a channel estimation by supposing a $\pi/2$ phase rotation at each symbol period and another one by supposing a $3\pi/8$ phase rotation. With the channel estimates that are obtained from the two computations, the most likely of the two results is used by the receiver to determine which modulation is used. A criterion such as the estimated *signal-to-noise ratio* (SNR) may be used, for example.

Of course, since a radio block comprises four bursts, each of these bursts uses the same modulation, and therefore four different blind detection operations may be performed by the receiver prior to the final decision on the modulation (depending on the implementation).

## 3.4 Case Studies

### 3.4.1 Generation of the Differential GMSK Signal

This section describes the different steps for the generation of the I and Q vector signals corresponding to the GMSK modulation. Refer to Section 3.1.1 for the definition of the GMSK modulation scheme.

The following operations are performed in the differential GMSK modulator, as represented in Figure 3.20:

- Differential Encoding

  Each data value $d_i = [0,1]$ is differentially encoded. The output of the differential encoder is:

  $$\hat{d}_i = d_i \oplus d_{i-1} \qquad \left(d_i \in \{0,1\}\right)$$

  where $\oplus$ denotes modulo 2 addition.

  The result is mapped onto $+1$ and $-1$ symbols, to form the modulating data value $\alpha_i$ input to the modulator, as follows:

  $$\alpha_i = 1 - 2\hat{d}_i \qquad \left(\alpha_i \in \{-1,+1\}\right)$$

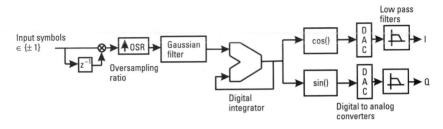

**Figure 3.20** Differential GMSK modulator synopsis.

- Filtering of the Input Bit Sequence by a Gaussian Filter

  The Gaussian filter is defined by its 3-dB bandwidth $B$, such that $BT = 0.3$, where $T$ is the symbol period. Since $T = 48/13 \, \mu s$, the filter bandwidth is 81.25 kHz. Theoretically, a Gaussian filter impulse response has an infinite duration. In practical realizations, this impulse response is truncated over a period of three to five symbol periods. This usually gives a satisfactory approximation of the ideal GMSK signal.

  The filtering of the input symbols may for instance be performed by convolution with FIR coefficients that implement the truncated Gaussian pulse. A possible solution consists of implementing this convolution in a lookup table. Indeed, since the Gaussian filter is truncated over $N$ symbols, there are $2^N$ different sequences at the output of the filter, depending on the $N$ input bits. It is therefore possible to perform the convolution calculations for each $N$-bit input sequence and to store the result in the table. Figure 3.21 shows an example of this look-up table, with a truncation of the Gaussian pulse over four symbol periods. Each output filter vector has a length of $4 \times OSR$, where $OSR$ is the oversampling ratio of the Gaussian pulse filter. This oversampling is required to achieve a good accuracy in the calculation of the modulated signal.

  Other design trade-offs concern the resolution of the pulse coefficients quantification in number of bits as well as the resolution of the multiplications and additions in the filtering process.

- Integration of the filtered output of the Gaussian filter. This integration can, for instance, be implemented with a digital accumulator.

- Generation of I and Q. The output of the integrator is the GMSK phase signal, $\varphi(t)$. The I and Q signals are obtained by calculating

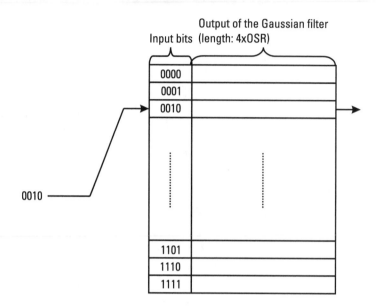

**Figure 3.21**   Example of Gaussian filter implemented by a look-up table.

cos φ(t), and sin φ(t). These calculations can, for instance, be implemented with tables that store the result of a cosine for different phase values (note that the sine table can be obtained from the cosine table with a π/2 phase shift). The number of bits used for the quantification of the output of these tables should be chosen carefully.

- Digital-to-analog conversion. The I and Q are converted into a continuous voltage signal. Again, the resolution of the converters in number of bits is to be considered. The conversion step is usually followed by a lowpass filter. The bandwidth and the group delay distortion of this filter may have an impact on the performance of the modulator.

Of course, other solutions are possible for the implementation of the GMSK baseband modulator. For instance, the phase signal could be completely kept in a look-up table. This way the use of the accumulator block for the integration of the instantaneous frequency can be avoided. All the parameters discussed for these different steps (length, OSR and resolution of the Gaussian impulse response, resolution of the additions and multiplications, resolution of the cosine table and DACs, distortion introduced by the low-pass analog filtering) have an influence on the phase error performance of the modulator, as well as on the generated spectrum due to modulation and wideband noise.

At the output of the modulator, the I and Q channels are produced, and they may be upconverted to RF through the RF transmitter.

### 3.4.2 Generation of the 8-PSK Signal

As seen in Section 3.1.2, the different steps of the modulated signal generation are the Gray mapping of the 3-bit words onto complex modulating symbols, the continuous $3\pi/8$ radian per symbol rotation, and the Gaussian filtering of the complex symbols represented by Dirac pulses at the symbol rate $1/T$ (the symbol duration is the same as the GMSK symbol duration, $48/13$ µs) by the linearized GMSK pulse $c_0(t)$.

A digital modulator for the 8-PSK I/Q generation is shown in Figure 3.22. In order to find the best trade-off between the different design constraints, the criteria that should be considered are the EVM and OOS performances (see Section 3.1.2.2), the spectrum due to modulation (Section 3.2.2), as well as the power versus time mask (given in Section 3.2.3).

The sources of imperfections that can degrade the modulation performance of the modulator are as follows:

- The resolution of the sine/cosine table of the digital modulator;
- The oversampling ratio used;
- The duration of the filter impulse response;
- The resolution of the multiplication-accumulations in the filtering process;
- The resolution of the filter coefficients;

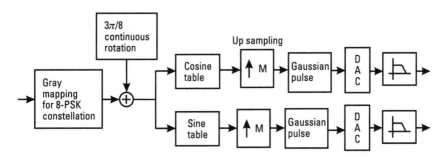

**Figure 3.22** Synopsis of the 8-PSK modulator.

- The resolution and the noise shape of the DACs;
- The analog lowpass filters' amplitude and group delay distortions.

Note that filtering may be implemented by a *finite impulse response* (FIR) filter structure, or by a look-up table.

### 3.4.3   RF Architecture Constraints of the EDGE Transmitter

The different transmitter architectures commonly in use for GSM/GPRS transmitters can be summarized as follows:

- *Intermediate frequency* (IF), for which the baseband signal is converted to an IF, and then to RF;
- Zero-IF or direct conversion, which consists of a single conversion stage from baseband to RF;
- Offset *phase locked loop* (PLL), or translation loop, based on the introduction of a phase modulating signal in a PLL system.

Except for the translation loop, which may be used only for *continuous phase modulations* (CPMs) (e.g., GMSK), the two other architectures can be implemented for 8-PSK.

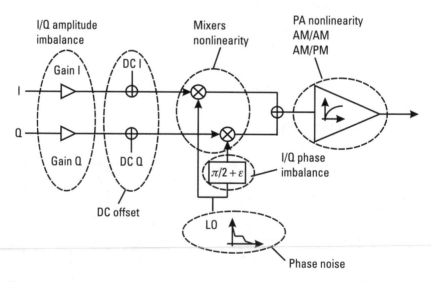

**Figure 3.23**  Zero-IF architecture RF impairments.

The different stages of the transmitter need to be dimensioned by considering the transmitter specifications: maximum output power, EVM and OOS, spectrum due to modulation and wideband noise, and power versus time template. Figure 3.23 shows as an example a generic zero-IF transmitter and the impairments usually associated with this type of RF architecture, which are as follows:

- On the baseband I and Q signals, there may be a dc offset due to the ADC and implementation of other analog functions.

- A mismatch in amplitude (the gain may be different on I and Q) and phase (the phase offset between both channels is not necessarily exactly $\pi/2$, but may be $\pi/2 \pm \varepsilon$).

- The phase noise due to the synthesizer.

- The PA is one of the most difficult functions in 8-PSK. Indeed, unlike for GMSK, the envelope of the RF signal is not constant. Therefore, any saturation in the PA affects the modulation quality. This is not the case with GMSK, where the amplitude distortion is not a problem, since the information is carried out by the phase of the carrier. The PA efficiency is better when it is used in its saturation region. This linearity constraint of the 8-PSK may have a significant impact on the MS power consumption, and this is why the 8-PSK maximum output power may be lower than for GMSK (see Section 3.2.1). Note that an implication of this is a reduced range for 8-PSK transmission compared to GMSK.

All these parameters are also to be considered for the design of an IF transmitter.

In addition to IF and zero-IF, there seems to be a tendency to develop so-called polar loop systems.

The idea is to use the polar coordinates of the modulated signal (angle and amplitude), instead of the I/Q vector coordinates [5].

This system can be divided into three parts (see Figure 3.24), as follows:

- One baseband processing function, whose purpose is to generate the modulated signal, as well as to perform the digital corrections that may be needed. The modulated signal is split into phase and amplitude signals.

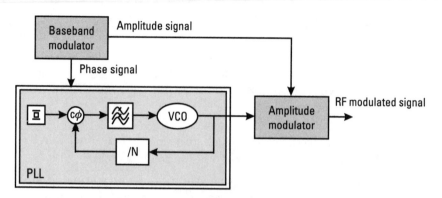

**Figure 3.24** Polar loop architecture for the EDGE 8-PSK modulation.

- The phase signal is used at the input of a PLL system (e.g., translation loop architecture or any improved version of this architecture), to allow for the phase modulation of the RF carrier.

- The PLL can provide only a constant envelope signal at the output of the VCO. Therefore, a third part is necessary to design a transmitter system that is 8-PSK capable, whose function is to generate the amplitude modulation signal at the RF output.

This kind of system needs careful synchronization between the phase and amplitude signals in order to generate an accurate modulation signal.

The advantage of this type of architecture is the cost compared to a zero-IF or to an IF transmitter. Indeed, it does not require any bulky filter, since the output noise of the PLL system is very low, because of the low-pass filtering effect of the loop. Besides, a good efficiency may be achieved with the amplitude modulation performed on the PA function.

### 3.4.4 GMSK Demodulation

The radio channels in mobile radio systems are usually distorted by multipath fading that is causing *intersymbol interference* (ISI) in the received signal. To remove this ISI, several types of equalizers may be used. The GSM standard gives requirements in terms of receiver performance, but it does not impose any special scheme for the demodulation. Many solutions to the demodulation problem for GSM channels are therefore possible. The most widely used equalizer, because of its rather low implementation cost and its optimal performance is the *maximum likelihood sequence estimator* (MLSE), based on the Viterbi algorithm. It is shown in [3], Chapter 4, that this algorithm can be used for the decoding of convolutional codes. It is also adapted to the equalization problem as described in this section.

First, it is important to introduce the linearization of the GMSK modulated signal. Although GMSK is a nonlinear modulation, it has been shown [2] that it can be approximated by a linear representation. Denoting $\{a_k\}$ as the symbols to be transmitted ($a_k = 1$ or $a_k = -1$), the complex GMSK signal at the output of the baseband modulator may be written:

$$S(t) = \sum_k j^k a_k p(t - kT) \tag{3.21}$$

where $p(t)$ is a real pulse given in [6], and $T$ is the symbol period. This shows that the differential GMSK modulator is equivalent to a system where the $\pm 1$ symbols are rotated by $\pi/2$ and filtered.

After transposition to RF and transmission through the channel, this signal is down converted to baseband and filtered by the receiver. The received baseband signal $r(t)$ can then be written as (see [7]):

$$r(t) = \sum_k a_k j^k h(t - kT) + n(t) \tag{3.22}$$

where $n(t)$ is the additive channel noise and interferences, $h(t)$ is the convolution of $p(t)$ with the global response of the channel comprising the transmitter filter and the multipath propagation channel. The noise is usually assumed to be white Gaussian. Sampling the signal at instants $t = iT$, (3.22) can be reformulated as

$$r_i = r(iT) = \sum_{k=0}^{+\infty} a_{i-k} h_k j^{i-k} + n_i = \sum_{k=0}^{+\infty} a_{i-k} h_k j^{i-k} e^{j(i-k)\frac{\pi}{2}} + n_i \tag{3.23}$$

In practice, the *channel impulse response* (CIR) is not assumed to be infinite, and a given symbol is affected by its $L - 1$ previous symbols, which leads to:

$$r_i = \sum_{k=0}^{L-1} a_{i-k} h_k j^{i-k} + n_i \tag{3.24}$$

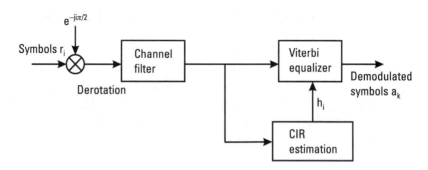

**Figure 3.25**  General synopsis of a GMSK demodulator.

These received signal samples are used for the different steps of the receiver.

Figure 3.25 gives a summary of the principal steps in a classical GMSK demodulator. A $\pi/2$ derotation is performed on the received sampled signal. This gives:

$$r'_i = r_i \cdot j^{-i} = \sum_{k=0}^{L-1} a_{i-k}\left( j^{-k} h_k \right) + \left( n_i j^{-i} \right) \tag{3.25}$$

so

$$r'_i = \sum_{k=0}^{L-1} a_{i-k} h'_k + n'_i \tag{3.26}$$

with

$$n'_i = n_i \cdot j^{-i} \text{ and } h'_k = h_k \cdot j^{-k}$$

Prior to the demodulation itself, the channel filter is intended to remove a part of the noise that is due to the receiver as well as the interference caused by the adjacent channels. Then, a CIR estimation is performed to provide the equalizer with the tap coefficients representing the propagation channel. The equalizer uses this information to remove the ISI due to the channel (multipath radio conditions, filtering in the transmitter and in the receiver) and provides the demodulated symbols.

The linear representation of the GMSK received signal, given in (3.26), is used for the computation of the CIR coefficients. In the GSM burst, the middamble bits have been chosen for this purpose.

The most basic solution for the CIR estimation is based on the correlation of the received middamble sequence with the known training sequence. The receiver evaluates the cross-correlation between the received signal corresponding to the training sequence and the known training sequence symbols. If we denote $(a_0, a_1, ..., a_{N-1})$ the $N = 16$ central symbols of the middamble, and $r = (r_0, r_1, ..., r_{N-1})$ the received symbols corresponding to these symbols, the following correlation may be calculated:

$$R_l = \frac{1}{N} \sum_{n=0}^{N-1} r'_{n+l} a_n \approx \frac{1}{N} \sum_{n=0}^{N-1} \sum_{k=0}^{L-1} a_n a_{n+l-k} h'_k \qquad (3.27)$$

Hence, $R_l$ is the correlation between $\{h'_k\}$ and $A_l = \frac{1}{N} \sum_{n=0}^{N-1} a_n a_{n+l}$ .

The training sequences have been chosen such that

$$A_l = \begin{cases} 1 \text{ if } l = 0 \\ 0 \text{ for } l \neq 0, |l| \leq 5 \end{cases} \qquad (3.28)$$

The training sequences are indeed chosen such that the auto-correlation of the 16 central bits is of course 1, and the inter-correlation for the 5 shifts on the left and right of the central sequence is 0. Such sequences are called *constant amplitude zero autocorrelation* (CAZAC). An example of training sequence is $(0,\underline{1},\underline{0},\underline{0},\underline{0},0,1,1,1,0,1,1,1,0,1,0,\mathbf{0},\mathbf{1},\mathbf{0},\mathbf{0},\mathbf{0},0,1,1,1,0)$. One can notice that the sequence is partially periodic: the 5 bits on the left correspond to the last bits of the central 16-bit sequence (underlined symbols) and the 5 bits on the right correspond to the beginning of the central sequence (bold symbols).

Therefore,

$$R_l = \sum_{k=0}^{L-1} h'_k A_{l-k} = h'_l \qquad (3.29)$$

Thus, by computing $R_l$ for different values of $l$, the receiver is able to make an estimation of the CIR coefficients (denoted $h'_k$), which are usually truncated over five or six symbols. In this approach, the sampled noise $\{n_k\}$ is simply neglected. Among the estimated CIR coefficients, a window of five to six successive estimates is therefore selected as the one that concentrates the greater energy.

The correlation between the received symbols and the known training sequence also allows the time synchronization of the receiver. Indeed, the position of the training sequence in the received sequence can be evaluated as the position that generates a peak in the correlation function. Note that other solutions for the channel estimation are possible.

**Note**   Since the receiver knows the training sequence, it is able to make an estimation of the noise samples, with the calculated CIR coefficients:

$$\hat{n}_i = r'_i - \sum_{k=0}^{L-1} a_{i-k} \hat{h}_k \qquad (3.30)$$

This information may be used by the receiver to estimate the SNR, since the noise power can be calculated, as well as the total received signal power. Indeed, given the information of the CIR, the receiver is able to make an estimation of the receiver noise. If we look at (3.30), knowledge of the taps $\{\hat{h}_k\}$, and of the transmitted symbols allows the estimation of the noise samples $n_i$, given the received samples $r'_i$. This is possible with the training sequence received samples, for instance, since the TSC is known to the receiver. Therefore, the noise power can be calculated, as well as the SNR.

Given (3.26), it is possible to estimate the maximum likelihood probability (see [8] for further details), which is proportional to

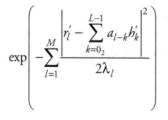

$$\exp\left( -\sum_{l=1}^{M} \frac{\left| r'_l - \sum_{k=0}^{L-1} a_{l-k} h'_k \right|^2}{2\lambda_l} \right)$$

where

$$\lambda_l = E\left[n_l n_l^*\right]$$

is the noise variance.

Thus, to perform the MLSE, the Euclidean distance $D$ between the received signal and the transmitted sequence must be minimized:

$$D^2 = \sum_{l=1}^{M} \left| r_l' - \sum_{k=-0}^{L-1} a_{l-k} h_k' \right|^2 \tag{3.31}$$

where $M$ is the size of the block to be demodulated.

In order to compute this distance, the receiver needs to know the CIR coefficients, $\{h_k\}$.

The implementation of the maximum likelihood equalizer requires the examination of all symbol sequences or paths through a trellis and the choice of the one that achieves the maximum of the likelihood function, which is the smallest Euclidean distance. The principle of the Viterbi algorithm, as described in [3], Section 4.3.2, is to avoid the estimation of the Euclidean distance for all the possible sequences of symbols. Rather, it is based on a trellis structure.

The number of states in the trellis is $2^{L-1}$, where $L$ is the number of taps of the CIR. An example, for 3 coefficients, is shown in Figure 3.26. This

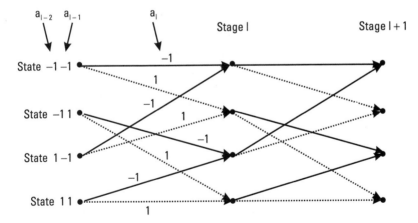

**Figure 3.26** Example of trellis diagram for a 3-tap Viterbi equalizer.

simple case assuming an ISI spread over 3 symbols is shown for illustrative purposes, and can easily be extended to additional coefficients. At a given stage l of the algorithm, the state represents the values of symbols $a_{l-1}$ and $a_{l-2}$, and the transition represents the symbol $a_l$. It is assumed that from each state, there is a branch corresponding to the symbol +1 and a branch corresponding to –1.

The Viterbi algorithm principle is summarized as follows:

- For each step $l$ = 0 to $M$
  — for each state $m$,
     estimation of the branch metrics between states $n$ and $m$,

$$\left( \delta_l \left( n,m \right) \right)$$

and between states $o$ and $m$

$$\left( \delta_l \left( o,m \right) \right):$$

$$\left| r'_l - \sum_{k=0}^{2} j^{l-k} a_{l-k} \hat{h}_k \right|^2$$

estimation of the path metric of state $m$, $M_l(m)$:

$$M_l \left( m \right) = \min \left( \delta_l \left( n,m \right) + M_{l-1} \left( n \right), \quad \delta_l \left( o,m \right) + M_{l-1} \left( o \right) \right)$$

  — The path that gives the highest state metric is discarded, the other is the survivor: At each step, half of the paths are eliminated.
- Trace-back: at the end of the equalization process, the best survivor among the paths that converge to the $2^{L-1}$ states is selected, and the demodulated sequence is deduced from all the branch labels of this path.

As seen in [3], the BER performance of the overall demodulating and decoding function may be improved with the concept of soft output equalizer. Indeed, at the output of the demodulator, the symbols are used for the convolutional decoding process (after de-interleaving).

The idea is not only to compute a maximum likelihood sequence estimation, but also to provide, for each symbol, an estimation of the probability that this symbol is the actual one that was transmitted. Instead of producing ±1 values, the equalizer output is a range of positive and negative values; the sign of these values makes it possible to determine whether the output symbol is +1 or −1; and the absolute value represents the reliability value. Doing this increases the correction capability of the subsequent Viterbi decoding algorithm. Several algorithms have been proposed for the estimation of the soft decisions. For example, a solution is proposed in [9].

**Note** The description of this section gives the digital receiver's general structure, which is simplified for the sake of clarity. In addition to the demodulation itself, a lot of preprocessing is required, such as frequency error correction, or dc offset suppression.

### 3.4.5 8-PSK Demodulation

For 8-PSK, the demodulation problem is more complicated. Indeed, the classical Viterbi implementation of the MLSE algorithm presented in the previous section requires much more processing power in the case of 8-PSK. This is because for an $M$-ary modulation, the number of states in the trellis is equal to $M^{L-1}$, for a CIR with $L$ coefficients. For instance, for $L = 5$ taps, the number of states is 16 in the case of GMSK, whereas it is 4096 in the case of 8-PSK, since there are eight possible transmitted values in 8-PSK and two in GMSK.

Instead of the Viterbi algorithm, a suboptimal technique may be implemented. Many equalization algorithms exist, and the description of all the possible solutions is well beyond the scope of this work. A reference on the subject is given at the end of the chapter (see [10]).

The following issues should be carefully considered when designing the EDGE demodulator, owing to the poorer performance of 8-PSK and its greater sensitivity to impairments:

- The distortion introduced in the receive channel filter has a significant impact on the performance, and a higher cutoff frequency compared with the GMSK receive filter may be needed.

- The receiver frequency error estimation must be able to cope with a varying frequency offset at each burst (refer to Section 3.3.2 describing the NER requirement).

- The dc offset due to the RF receiver, particularly in the case of the zero-IF receiver (see [3], Section 4.3.4.2) has a greater effect on receiver performance than in the GMSK case. Enhanced dc estimation techniques may be needed (see [11] for more information).

# References

[1] Murota, K., and K. Hirade, "GMSK Modulation for Digital Mobile Radio Telephony," *IEEE Trans. on Communications*, Vol. 29, July 1981, pp. 1044–1050.

[2] Laurent, P. A., "Exact and Approximate Construction of Digital Phase Modulation by Superposition of Amplitude Modulated Pulses," *IEEE Trans. on Communications*, Vol. 34, No. 2, February 1986.

[3] Seurre, E., P. Savelli, and P. J. Pietri, *GPRS for Mobile Internet*, Norwood, MA: Artech House, 2003.

[4] 3GPP TS 05.05 Radio Transmission and Reception (R99).

[5] Heimbach, M., "Polarizing RF Transmitters for Multimode Operation," *Communication Systems Design*, October 2001.

[6] Kawas Kaleh, G., "Differentially Coherent Detection of Binary Partial Response Continuous Phase Modulation with Index 0.5," *IEEE Trans. on Communications*, Vol. 39, No. 9, September 1991.

[7] Benelli, G., A. Garzelli, and F. Salvi, "Simplified Viterbi Processors for the GSM Pan-European Cellular Communication System," *IEEE Trans. on Vehicular Technology*, Vol. 43, No. 4, November 1994, pp. 870–877.

[8] Benelli, G., et al., "Design of a Digital MLSE Receiver for Mobile Radio Communications," *Globecom'91*, Phoenix, AZ, December 1991, pp. 1469–1473.

[9] Hagenauer, J., and P. Hoeher, "A Viterbi Algorithm with Soft Decision Outputs and Its Applications," *Proc. Globecome '89*, Dallas, TX, November 1989, pp. 47.1.1–47.1.7.

[10] Gerstacker, W. H., and R. Schober, "Equalization Concepts for EDGE," *IEEE Trans. on Wireless Communications*, Vol. 1, No. 1, January 2002.

[11] Krakovsky, C., W. Xu, and F. von Bergen, "Joint Channel and DC Offset Estimation and Synchronization with Reduced Computational Complexity for an EDGE Receiver," *VTC Conference*, 2001.

# 4

# Physical Link Layer

This section describes the EGPRS-specific concepts that are used for the management of the link quality control. These concepts deal with physical layer aspects as well as RLC layer aspects. In fact, some of these concepts overlaps each other and cannot be explained separately. First, the way the coding is performed for the MCSs families is explained. The second part deals with the link adaptation scheme and, among other things, the IR mechanism. The last section gives some case studies related to link adaptation.

## 4.1 Channel Coding

As described previously, EGPRS relies on nine MCSs that are used to carry data. Four of them are associated with GMSK modulation (MCS-1 to MCS-4), and the others are associated with 8-PSK modulation (MCS-5 to MCS-9). EGPRS CSs based on GMSK are different from GPRS CSs. Moreover, in the EGPRS mode, data is transferred over PDTCH, using the MCSs only; the GPRS CSs are not used at all. However, on the signaling channels, EGPRS uses the same coding as GPRS (CS-1).

### 4.1.1 Channel Coding for EGPRS PDTCH

The RLC/MAC blocks for EGPRS data transfer are used as transport units on the air interface. One EGPRS RLC/MAC block is carried by one radio block that is transmitted over four consecutive bursts on four TDMA frames on a given PDCH.

EGPRS RLC/MAC blocks are composed of a header part and one or two data parts. Three header formats have been defined. The header format depends on the MCS used. Unlike GPRS, the header and the data parts of the block are not coded with the same protection level. As explained in Section 1.3.1.4, a GPRS receiver detects which CS is used for a radio block in downlink or uplink, thanks to the stealing flag. For EGPRS, even more information is needed to decode a radio block—specifically, which modulation and which MCS was used as well as which puncturing pattern (there are several patterns for a given MCS).

The receiver must first perform a blind detection of the modulation (8-PSK or GMSK) and then rely on the stealing flag to identify the type of header that was used by the transmitter; eventually, it must decode the header, which holds the rest of the information about the radio block (MCS number and puncturing pattern).

There is a common header for MCS-1, MCS-2, MCS-3, and MCS-4; a common header for MCS-5 and MCS-6; and another one for MCS-7, MCS-8, and MCS-9. The header type is indicated by the stealing flag combinations described in Table 4.1.

For GMSK MCSs, there is also an extra stealing flag made of four bits (0000), which could be used for future header formats (other CSs than MCS1, MCS2, MCS3, and MCS4 could be introduced in the future).

The header is always coded with a robust coding, ensuring good probability of decoding in all radio conditions. The header decoding is required to decode the data and to perform IR. This will be explained later in this chapter. It is to be noted that a given MCS format is slightly different in uplink

**Table 4.1**
Mapping Between Header Types and Stealing Flag

| Header Type | Stealing Flag Pattern | Modulation |
|---|---|---|
| MCS-1, MCS-2, MCS-3, MCS-4 | 00010110 | GMSK |
| MCS-5, MCS-6 | 00000000 | 8-PSK |
| MCS-7, MCS-8, MCS-9 | 11100111 | 8-PSK |

and in downlink, owing to the different header formats. The MCSs presented in this section concern only the downlink, for the purpose of providing an example.

### 4.1.1.1 MCS-1 to MCS-4

One of the main requirements that were agreed upon for the design of the EGPRS system was the possibility to multiplex GPRS and EGPRS users on the same PDCH in both uplink and downlink directions. That means a "GPRS only" mobile shall be able to decode correctly its USF in downlink blocks addressed to an EGPRS mobile when the network wants to allocate it one uplink occurrence. On the other hand, an EGPRS mobile shall also be able to read the USF inside blocks addressed to a GPRS or an EGPRS mobile.

When 8-PSK modulation is used for downlink transmissions to an EGPRS mobile, a "GPRS only" mobile will definitively not be able to decode the USF. The only possibility to allocate one uplink occurrence to a GPRS user is to transmit the downlink block that contains the USF using the GSMK modulation. Therefore, for radio blocks transmitted with the GMSK modulation (i.e., with MCS-1 to MCS-4), it was decided to reuse the same USF coding mechanism as in GPRS.

In GPRS, the USF coding is different depending on the CS used. For CS-1 to CS-3, the USF is precoded with a block code and then encoded together with the data with a convolutional code. For CS-4, the USF is preencoded only with a block code that is very easy to decode. In spite of the different coding for CS-2, CS-3, and CS-4, the USF pattern obtained after encoding is the same. The same USF coding as for CS-2, CS-3, and CS-4 is used for the GMSK-modulated EGPRS radio blocks. It was also decided to choose the same stealing flags as CS-4 to identify MCS-1 to MCS-4. A legacy GPRS MS will first detect the stealing flag, then decode the USF; it will not succeed in decoding the data part (parity check will fail) if it is encoded with any of the MCSs. This mechanism provides a mean to multiplex EGPRS and GPRS mobiles in uplink.

Figure 4.1 shows the coding principle for MCS-1 to MCS-4. On the bottom part of the figure, the four NBs that compose the radio block are represented. Each one is transmitted in four consecutive TDMA frames. The NB comprises in the middle a *training sequence* (TS) and on each side two stealing bits and two data sequences. Note that four extra stealing bits are spread over the four bursts.

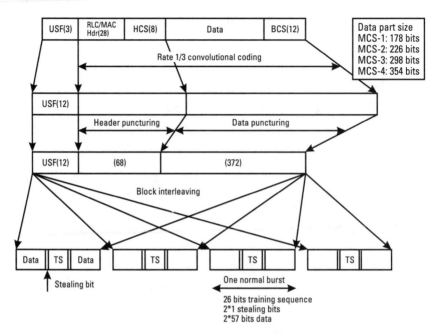

**Figure 4.1**    Coding for MCS-1 to MCS-4 (downlink direction).

The input used for MCS1 to MCS4 encoding is made of an RLC/MAC header part and a data part. The header size is the same for all the GMSK MCSs, whereas the data size increases with the MCS number. An 8-bit *header check sequence* (HCS) is added to the header, and a 12-bit BCS to the data part. These parity bits are used for error detection.

The coding principle is almost the same for all the MCSs. The first step (for downlink only) consists of precoding the USF (12 coded bits from 3 information bits). The next step is the header coding (including HCS) which consists in a "cyclical" encoding with a rate 1/3 convolutional code ("tail-biting" technique). This code is used for error correction. Then the header is lightly punctured.

The data (including BCS) is then encoded by adding 6 tail bits and using the same rate 1/3 convolutional code. Note that this convolutional code (the mother code) is the same for all the MCSs. The polynomials of this convolutional code are given in Table 4.2.

The puncturing of the data part is variable and depends on the MCS used. However, for the same MCS, several puncturing schemes are available and can be used by the transmitter.

**Table 4.2**
Convolutional Code Parameters for all the MCSs

| Modulation and Coding Schemes | Polynomials | Code Rate |
|---|---|---|
| All MCSs | $G4 = D^6+D^5+D^3+D^2+1$ | 1/3 |
| | $G5 = D^6+D^4+D+1$ | |
| | $G7 = D^6+D^3+D^2+D+1$ | |

As a matter of fact, two puncturing schemes exist for MCS-1 and MCS-2. For MCS-3 and MCS-4, three puncturing schemes have been defined. The use of these different puncturing schemes for the same MCS is explained in Section 4.2. Which MCS and puncturing scheme was used is indicated in the header of the RLC/MAC block by the field CPS (see Section 2.3.5.1).

After puncturing, no matter which MCS is used, there are always 452 (encoded) bits left. The last step consists of "block interleaving" these 452 bits (plus 8 bits of the stealing flag and 4 bits of the extra stealing flag) over the four normal bursts that compose the radio block. The header is always interleaved over the four bursts.

### 4.1.1.2 MCS-5 and MCS-6

MCS-5 and MCS-6 are associated with 8-PSK modulation. The encoding principle is almost the same as for MCS-1 to MCS-4.

The input that is received before channel encoding is composed of a header and a data part. The data part can have two different lengths, depending on whether the block is encoded with MCS-5 or MCS-6 (the header is the same for both). The coding principle is described in Figure 4.2. The only differences compared with the GMSK MCSs concerns the USF, which is precoded with 36 bits, and the RLC/MAC header, which is not punctured after the convolutional coding. Two puncturing schemes for the data part are defined for each of the two MCSs. The number of bits available at the end of the encoding is three times higher than the GMSK MCSs due to the use of 8-PSK modulation for their transmission $[1,384 + 8 = 3 * (452 + 8 + 4)]$.

The encoded data is then interleaved over the four bursts.

As stated in Chapter 3, each state of the 8-PSK constellation represents three bits. But if we look at the distribution of the bits, it appears that for each symbol the error probability of the two first bits is lower than the one of the third bit. The reason is quite simple; let's assume we want to transmit the

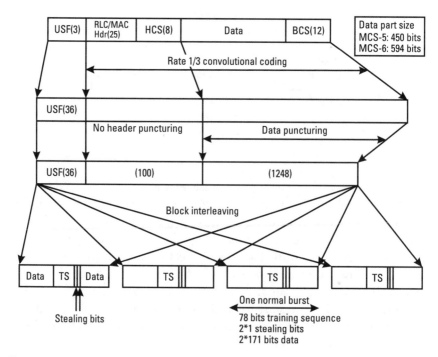

**Figure 4.2**    Coding for MCS-5 and MCS-6 (downlink direction).

symbol labeled with the bits "000." For the first and second bit, the value "0" will be well decoded in one entire half plane. For the third bit, the plane is split into four areas leading to a higher error probability. It is actually the same for any symbol of the constellation.

As the third bit has a higher error probability and correct header decoding is mandatory to be able to decode the data, it was decided to avoid putting bits of the header on every third position: The interleaving scheme is such that it spreads the header bits on the first two positions only. This way, the robustness of the RLC/MAC header decoding is increased. This is also the reason why the stealing bits are not positioned on each side of the TS (as in GPRS) but are grouped on the same side, using only the first two positions of a symbol.

### 4.1.1.3    MCS-7, MCS-8, and MCS-9

MCS-7, MCS-8, and MCS-9 have the capability to include two data parts within the same radio block (see Figure 4.3). For these MCSs, the input is composed of a RLC/MAC header, the data part of the first RLC data block,

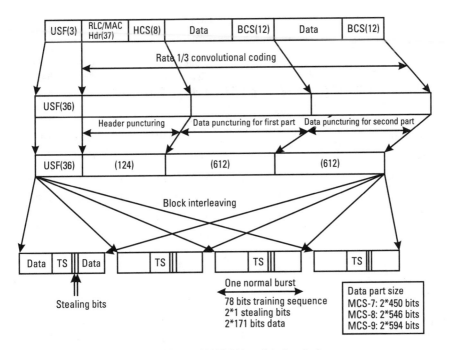

**Figure 4.3** Coding for MCS-7, MCS-8, and MCS-9 (downlink direction).

and the data part of the second RLC data block. A parity check sequence is added to the different parts that compose the input (8 bits for the header and 12 bits for each data part). The header and the data parts are coded by using the rate 1/3 convolutional code and are then punctured.

For each data part the receiver can use three puncturing schemes. Note that different puncturing schemes can be used for each data part. The MCS and the associated puncturing schemes used to encode the two data parts are indicated within the header.

The 1,384 bits obtained (plus 8 stealing bits) are interleaved over the four bursts. For MCS-7, each data part is interleaved over the four bursts. For MCS-8 and MCS-9, each data part is interleaved over two bursts. Upon reception of a radio block encoded with one of these MCSs, it may happen that one data part is decoded and not the other (assuming correct decoding of the header).

### 4.1.1.4 Summary of the Different Coding Parameters

Table 4.3 details the different parameters of the MCSs for the uplink and downlink directions.

**Table 4.3**
MCS Parameters

| MCS | USF | Precoded USF | Header | HCS | Header Code Rate | Data | BCS | Data Code Rate | Data Rate in Kbps |
|---|---|---|---|---|---|---|---|---|---|
| MCS-1 DL | 3 | 12 | 28 | 8 | ≈1/2 | 178 | 12 | ≈1/2 | 8.9 |
| MCS-1 UL | | | 31 | 8 | ≈1/2 | 178 | 12 | ≈1/2 | 8.9 |
| MCS-2 DL | 3 | 12 | 28 | 8 | ≈1/2 | 226 | 12 | 0.64 | 11.3 |
| MCS-2 UL | | | 31 | 8 | ≈1/2 | 226 | 12 | 0.64 | 11.3 |
| MCS-3 DL | 3 | 12 | 28 | 8 | ≈1/2 | 298 | 12 | 0.83 | 14.9 |
| MCS-3 UL | | | 31 | 8 | ≈1/2 | 298 | 12 | 0.83 | 14.9 |
| MCS-4 DL | 3 | 12 | 28 | 8 | ≈1/2 | 354 | 12 | ≈1 | 17.7 |
| MCS-4 UL | | | 31 | 8 | ≈1/2 | 354 | 12 | ≈1 | 17.7 |
| MCS-5 DL | 3 | 36 | 25 | 8 | ≈1/3 | 450 | 12 | ≈1/3 | 22.5 |
| MCS-5 UL | | | 37 | 8 | ≈1/3 | 450 | 12 | ≈1/3 | 22.5 |
| MCS-6 DL | 3 | 36 | 25 | 8 | ≈1/3 | 594 | 12 | ≈1/2 | 29.7 |
| MCS-6 UL | | | 37 | 8 | ≈1/3 | 594 | 12 | ≈1/2 | 29.7 |
| MCS-7 DL | 3 | 36 | 37 | 8 | ≈1/3 | 2*450 | 12 | ≈3/4 | 45 |
| MCS-7 UL | | | 46 | 8 | ≈1/3 | 2*450 | 12 | ≈3/4 | 45 |
| MCS-8 DL | 3 | 36 | 37 | 8 | ≈1/3 | 2*546 | 12 | 0.90 | 54.6 |
| MCS-8 UL | | | 46 | 8 | ≈1/3 | 2*546 | 12 | 0.90 | 54.6 |
| MCS-9 DL | 3 | 36 | 37 | 8 | ≈1/3 | 2*594 | 12 | ≈1 | 59.4 |
| MCS-9 UL | | | 46 | 8 | ≈1/3 | 2*594 | 12 | ≈1 | 59.4 |

## 4.1.2 Channel Coding for the Other Channels

In order to keep compatibility with the GPRS system for signaling phases, EGPRS uses exactly the same channel coding on the other channels than PDTCH. All the signaling on PACCH, PBCCH, PCCCH uses CS-1, defined for GPRS. The RACH and PRACH use the same channel coding, but two new training sequences have been defined and can be used only by the EDGE mobile. Their use is detailed in the following chapter.

## 4.2 Link Quality Control

The link quality control mechanism manages the choice of coding for the transmission of the radio blocks on the radio interface. It evaluates the radio conditions based on quality measurements and adapts the channel coding in accordance with them. The mechanism has been enhanced compared with GPRS because of the addition of several link management schemes.

The first improvement concerns the quality measurements and their averaging. For GPRS, the measurements were not well suited to packet transfer and to the different CSs (see Section 2.3.3.3). Moreover when receiving CS-4, the mobile was not required to perform quality measurements. In EGPRS, the link adaptation algorithm that manages the choice of the MCS to be used for uplink and downlink transmission uses these new measurements. The link adaptation mechanism is shared between the mobile and the network.

The second enhancement is the introduction of IR (also known as hybrid type II ARQ) when the mobile operates in RLC acknowledged mode. This mechanism provides a lot of flexibility in managing the MCSs and a lot of safety in the choice of the MCS.

### 4.2.1 Measurements for Link Quality Control

In order to provide a reliable estimation of the link quality, a new metric was introduced. This metric is the BEP, which is estimated on a burst-by-burst basis before channel decoding. It is used to derive two statistical parameters that are used by the link quality control algorithm at the network side. These statistical parameters are called MEAN_BEP and CV_BEP. Note that the mobile in downlink takes into account only the quality measurements made on radio blocks that are addressed to it (the header shall be correctly decoded).

#### 4.2.1.1 MEAN_BEP Computation

The BEP that is calculated on a burst-by-burst basis is then averaged on a radio block basis (four bursts); a mean value is then obtained. This average is called mean(BEP).

$$mean(BEP) = \frac{1}{4} \cdot \sum_{i=1}^{4} BEP_i \qquad (4.1)$$

In the MS, the mean(BEP) values are filtered on a time slot and modulation basis (MEAN_BEP_TS); depending on the modulation that was used in downlink, the same BEP value does not reflect the same downlink quality. So the mobile has to maintain two MEAN_BEP_TS values per time slot: one for GMSK and one for 8-PSK.

For a given modulation, the MEAN_BEP is then obtained by averaging the corresponding MEAN_BEP_TS values for all the allocated downlink time slots. Depending on the network reporting order, the MS reports the two MEAN_BEP values (one for each modulation), or the two MEAN_BEP values and the two MEAN_BEP_TS values for each assigned time slot. Table 4.4 shows the mapping between the MEAN_BEP value and the reported value.

**Table 4.4**
Range of Reported Parameters MEAN_BEP and MEAN_BEP_TS

| Reported MEAN_BEP | Range of log10 (Actual BEP) |
|---|---|
| MEAN_BEP_0 | > −0.60 |
| MEAN_BEP_1 | −0.70 to −0.60 |
| MEAN_BEP_2 | −0.80 to −0.70 |
| MEAN_BEP_3 | −0.90 to −0.80 |
| MEAN_BEP_4 | −1.00 to −0.90 |
| MEAN_BEP_5 | −1.10 to −1.00 |
| MEAN_BEP_6 | −1.20 to −1.10 |
| MEAN_BEP_7 | −1.30 to −1.20 |
| MEAN_BEP_8 | −1.40 to −1.30 |
| MEAN_BEP_9 | −1.50 to −1.40 |
| MEAN_BEP_10 | −1.60 to −1.50 |
| MEAN_BEP_11 | −1.70 to −1.60 |
| MEAN_BEP_12 | −1.80 to −1.70 |
| MEAN_BEP_13 | −1.90 to −1.80 |
| MEAN_BEP_14 | −2.00 to −1.90 |
| MEAN_BEP_15 | −2.10 to −2.00 |

*From:* [1].

**Table 4.4** (continued)

| Reported MEAN_BEP | Range of log10 (Actual BEP) |
|---|---|
| MEAN_BEP_16 | −2.20 to −2.10 |
| MEAN_BEP_17 | −2.30 to −2.20 |
| MEAN_BEP_18 | −2.40 to −2.30 |
| MEAN_BEP_19 | −2.50 to −2.40 |
| MEAN_BEP_20 | −2.60 to −2.50 |
| MEAN_BEP_21 | −2.70 to −2.60 |
| MEAN_BEP_22 | −2.80 to −2.70 |
| MEAN_BEP_23 | −2.90 to −2.80 |
| MEAN_BEP_24 | −3.00 to −2.90 |
| MEAN_BEP_25 | −3.10 to −3.00 |
| MEAN_BEP_26 | −3.20 to −3.10 |
| MEAN_BEP_27 | −3.30 to −3.20 |
| MEAN_BEP_28 | −3.40 to −3.30 |
| MEAN_BEP_29 | −3.50 to −3.40 |
| MEAN_BEP_30 | −3.60 to −3.50 |
| MEAN_BEP_31 | < −3.60 |

*From:* [1].

MEAN_BEP_TS is averaged by means of the averaging filter:

$$MEAN\_BEP\_TS_n = [1 - a(t)] \cdot MEAN\_BEP\_TS_{n-1} + a(t) * mean(BEP)_n \qquad (4.2)$$

$n$ is the iteration index and $a(t)$ is the forgetting factor. One can see that the forgetting factor—unlike GPRS—is dependent on the time that elapsed between two-block receptions. As several mobiles can be multiplexed on the same time slot, it can happen that during some block periods, the network does not address any block to a particular MS. The averaging process takes into account this phenomenon. When the time between two transmissions to the same mobile increases, the weight that is put on the last measurement is increased, since it represents better the actual radio channel quality. The exact computation of $a(t)$ is not given here but can be found in [1].

The MEAN_BEP is then obtained (independently for each modulation) with the following formula:

$$MEAN\_BEP = \frac{\sum_{TS} b(t)_{TS} * MEAN\_BEP\_TS}{\sum_{TS} b(t)_{TS}} \qquad (4.3)$$

where $b(t)_{TS}$ traduces the reliability (time dependent) of the average MEAN_BEP_TS compared with the same measure on the other time slots. On the network side, there are no particular requirements on the averaging of the mean(BEP). Any averaging process can be implemented.

### 4.2.1.2 CV_BEP Computation

On a block-by-block basis, the MS and the network evaluate the coefficient of variation of the channel quality CV(BEP) that is deduced from the mean(BEP) by

$$CV(BEP) = \frac{std(BEP)}{mean(BEP)} \qquad (4.4)$$

where $std(BEP)$ is the standard deviation of the four BEP estimates within the radio block.

$$std(BEP) = \sqrt{\frac{1}{3} \cdot \sum_{i=1}^{4} \left( BEP_i - mean(BEP) \right)^2} \qquad (4.5)$$

On the mobile side, the CV(BEP) is averaged on a time slot and modulation basis (CV_BEP_TS). The reported value CV_BEP to the network is averaged on all the allocated time slots. Table 4.5 shows the mapping of the CV_BEP value on the different reported value CV_BEP_X. The mobile performs two computations of CV_BEP value, one for GMSK and another for 8-PSK.

CV_BEP_TS is averaged by means of the averaging filter:

$$CV\_BEP\_TS_n = [1 - a(t)]CV\_BEP\_TS_{n-1} + a(t) * CV(BEP)_n \tag{4.6}$$

$n$ is the iteration index and $a(t)$ is the forgetting factor.

The CV_BEP is then obtained with the following formula:

$$CV\_BEP = \frac{\sum_{TS} b(t)_{TS} * CV\_BEP\_TS}{\sum_{TS} b(t)_{TS}} \tag{4.7}$$

On the network side, there is no requirement to average the CV(BEP) values.

The first parameter, MEAN_BEP, describes the impact of the received C/I, the time dispersion, and the velocity. The second parameter CV_BEP describes how the channel quality varies from burst to burst and can hence be used to reflect interleaving gain/loss caused by the velocity and *frequency hopping* (FH).

**Table 4.5**
Mapping Table for Coefficient of Variation

| Reported CV_BEP | Range of CV_BEP |
|---|---|
| CV_BEP_0 | 2.00 – 1.75 |
| CV_BEP_1 | 1.75 – 1.50 |
| CV_BEP_2 | 1.50 – 1.25 |
| CV_BEP_3 | 1.25 – 1.00 |
| CV_BEP_4 | 1.00 – 0.75 |
| CV_BEP_5 | 0.75 – 0.50 |
| CV_BEP_6 | 0.50 – 0.25 |
| CV_BEP_7 | 0.25 – 0.00 |

*From:* [1].

## 4.2.2  IR Mechanism

IR consists of sending $n$ times the same block until the block is decoded. The soft values of the previous unsuccessful transmissions are used. The coding rate is decreased at each transmission, increasing the probability of successful decoding. IR allows the reduction of the coding rate with the retransmission of the same block. This mechanism is used only when the RLC Protocol operates in acknowledged mode. In Section 4.2.2.1, the principle of IR is described. In the second, the way it is managed during data transfer in EGPRS mode is described.

### 4.2.2.1   IR Principle

The principle of IR is very simple. It consists of using the information that was sent in the previous transmissions of the same block to increase the probability of decoding. Once a data block has been transmitted, if the receiver did not decode it correctly, it stores the soft values at the output of the demodulator in order to reuse them after the retransmission of the block. When the transmitter sends this block again, it uses a different puncturing scheme so that the bits that have been transmitted the second time do not carry the same information as the first time. If the puncturing scheme from the first block is completely disjoint from the second one, the number of coded bits that have been transmitted after the second transmission is twice as high. Thus, the coding rate has been divided by two. After the third transmission, the coding rate has been divided by three. The probability to decode the block is then increased after each retransmission.

**Note**   If the puncturing schemes are not disjoint, soft values corresponding to the same bit are added. The coding rate is the same, but the decisions' average reliability increases: after combining, the magnitude of same sign decisions increases, while the magnitude of opposite sign decisions (thus unreliable) decreases.

Figure 4.4 illustrates of how IR runs from a transmitter point of view and from a receiver point of view. Suppose a data block containing $n$ information bits has to be sent OTA interface. The data block is encoded using a convolutional code with a code rate 1/3. This leads to $3*n$ coded bits. In order to increase the coding rate, the encoded data are punctured (using one puncturing scheme among the three available). Let us suppose that two-thirds of the bits are punctured by the puncturing schemes. There are $n$ bits left, and the coding rate is then equal to 1.

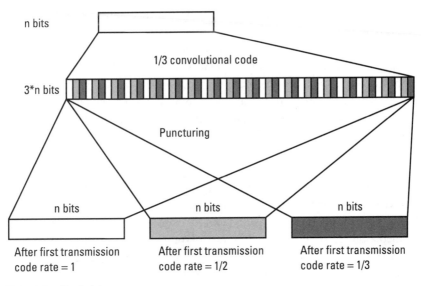

n bits

1/3 convolutional code

3*n bits

Puncturing

n bits          n bits          n bits

After first transmission    After first transmission    After first transmission
code rate = 1               code rate = 1/2             code rate = 1/3

**Figure 4.4**    IR principle.

The three puncturing schemes are defined in order to have no common bit. For the first transmission, the bits that are transmitted are the white ones in the figure. For the second transmission, the bits in light gray are transmitted, and for the third, the bits in dark gray. At each transmission, $n$ coded bits are transmitted on the air interface.

Upon the first reception, the receiver that knows the punctured positions fills them with zeroes before decoding (erasure). If the receiver decodes the block correctly, the transmission will have been performed with a coding rate equal to 1. If the block is not decoded, the receiver stores in its internal memory the soft decisions corresponding to the received bits.

Upon the second reception, the receiver fills one half of the punctured positions with zeroes and the other half with the soft values from the first transmission before decoding. Whether the block has been correctly decoded or not, the transmission will have been performed with a coding rate equal to 1/2: $2*n$ coded bits have been transmitted for $n$ information bits.

Upon the third reception, the receiver fills all the punctured positions with the received soft values and those of the previous transmissions. The coding rate is then 1/3. This iterative process can be performed until the block is decoded. However when all the bits have been sent with the various puncturing schemes, the gain provided by the next transmissions is limited.

The big constraint that is brought by the IR is linked to the memory that is required for the storage of the soft decisions. At each iteration, the soft decisions have to be stored. After each transmission, the storage of the soft decisions increases the amount of memory. In order to keep the price of the mobile as low as possible, it is necessary to limit the memory that is used for IR. However, in order to ensure a minimum service quality, a test was defined so that a minimum amount of memory is available in the MS. This test is described in Section 4.3.1.

### 4.2.2.2   IR Management

IR is optional for the BTS but is mandatory for the MS. For EGPRS, IR was introduced by means of puncturing schemes that were defined for each MCS. Depending on the MCS, two or three puncturing schemes were defined. Table 4.6 gives the number of puncturing schemes defined per MCS. The same names have been used for the puncturing schemes of all the MCSs, but they are different for each MCS.

The transmitter behavior is very simple: after each transmission of the same block, it changes the puncturing scheme that is used. For the first transmission P1 is used, then P2, and eventually P3 or P1 if it does not exist. The MCS and puncturing scheme that were used are indicated within the CPS field of the RLC/MAC header of the radio block.

From the receiver point of view, things are more complex: The receiver has to manage and store all but not yet successfully decoded radio blocks that

**Table 4.6**
Puncturing Schemes for the Various MCSs

| MCS | Puncturing Schemes |
| --- | --- |
| MCS-1 | P1, P2 |
| MCS-2 | P1, P2 |
| MCS-3 | P1, P2, P3 |
| MCS-4 | P1, P2, P3 |
| MCS-5 | P1, P2 |
| MCS-6 | P1, P2 |
| MCS-7 | P1, P2, P3 |
| MCS-8 | P1, P2, P3 |
| MCS-9 | P1, P2, P3 |

are addressed to it. It should be noted that if the header of the block is not decoded, the mobile is unable to identify whether the block is addressed to it or to another mobile, and in this case, IR provides no gain. These soft values cannot be reused. Moreover, the decoding of the header is necessary to identify the MCS and the puncturing scheme that were used to send the radio block. That is the reason why the header was encoded with high protection so that the probability of being able to decode the header is very high whatever the radio conditions.

The receiver has first to decode the header. Once the header has been decoded, the TFI and the BSN identify the RLC data block. If the received RLC data block is a retransmission, the receiver performs a soft combining of the soft values from this retransmission and the previous ones. Then channel decoding can be performed.

For MCS-7, MCS-8, and MCS-9, there are two RLC data parts per radio block. However, they are encoded separately. From an IR point of view they are managed as if they had been received from two different radio blocks.

Table 4.7 gives the coding rates achieved after each new transmission of the same block using a new puncturing scheme. Note that the puncturing schemes are not always disjoint. For example, after the first transmission using MCS-1, the coding rate is 1/2 and goes down to 1/3 after the second

**Table 4.7**
Coding Rates Achieved After $n$ Transmissions of the Same Block for the Various MCSs

| MCS | Coding Rate After First Transmission | Coding Rate After Second Transmission | Coding Rate After Third Transmission |
|---|---|---|---|
| MCS-1 | 1/2 | 1/3 | |
| MCS-2 | 0.64 | 1/3 | |
| MCS-3 | 0.83 | 0.42 | 1/3 |
| MCS-4 | 1 | 1/2 | 1/3 |
| MCS-5 | 0.32 | 1/3 | |
| MCS-6 | 1/2 | 1/3 | |
| MCS-7 | 3/4 | 3/8 | 1/3 |
| MCS-8 | 0.9 | 0.45 | 1/3 |
| MCS-9 | 1 | 1/2 | 1/3 |

transmission. If the two puncturing schemes had been disjoint, the coding rate would have been 1/4 after the second transmission.

### 4.2.3  Link Adaptation Mechanism

The principle of link adaptation is to adapt the modulation and CS to the radio conditions. When the radio conditions are poor, a MCS with a low coding rate is chosen, leading to a lower throughput. When the radio conditions are very good, a high coding rate is chosen, leading to higher throughput. During the data transfer, the network evaluates the link quality and decides which MCS to use accordingly.

EGPRS uses a different mechanism that allows a more efficient adaptation of the link depending on the radio conditions. The transfer of RLC data blocks in the acknowledged RLC/MAC mode can be controlled by a selective type I ARQ mechanism or by a selective type II hybrid ARQ (IR) mechanism within one TBF. The mechanisms that are described in this section are valid only for operation in RLC acknowledged mode.

The link adaptation scheme relies on the MCS family concept that was introduced in Chapter 2.

#### 4.2.3.1  Segmentation Mechanism

As explained in Chapter 2, the EGPRS MCSs have been designed in such a way that an RLC data block that is transmitted with a high coding rate can be retransmitted with a lower coding rate if the receiver does not decode it correctly. This is possible because of the MCS family concept that is described in Section 2.3.3.1. The MCSs are divided into four families. A radio block encoded with a given MCS carries a data payload whose length is a multiple of the basic data unit of its family. Whenever one radio block carrying multiple basic data units is not decoded correctly, one payload data unit can be retransmitted separately in another radio block, with a different MCS belonging to the same family, with a lower coding rate.

MCS-7, MCS-8, and MCS-9 carry two RLC data parts that are identified by two BSNs within one radio block. Each RLC data part identified by a BSN can be retransmitted later in one radio block.

Figure 4.5 shows the case of two blocks transmitted with MCS-9 that are retransmitted later within two MCS-6 radio blocks.

The previous principle consists of retransmitting two different RLC data blocks within two radio blocks, although they were transmitted within one radio block the first time. However it is also possible to split one RLC

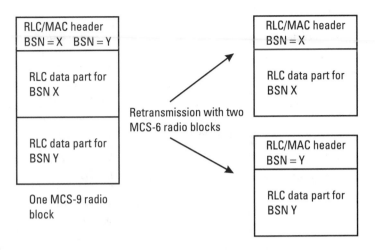

**Figure 4.5** Retransmission of RLC data blocks.

data block into two parts and to send it in two different radio blocks. This is called the segmentation mechanism. This is the case for an MCS-4, MCS-5, and MCS-6 encoded RLC data block that can be retransmitted, respectively, by two MCS-1, MCS-2, and MCS-3 radio blocks. In this case, as the RLC data part of the original block is split into two radio blocks, the header of each of these two radio blocks will contain the same BSN. In order to identify which radio block contains the first segment of the RLC data part, the transmitter uses the SPB field of the header. Figure 4.6 illustrates this mechanism.

The network indicates to the mobile whether or not it is allowed to use the segmentation mechanism in the uplink direction during the establishment phase of the uplink TBF. When the IR is used in the uplink direction, there is no need to use the segmentation mechanism. In fact, as explained in the previous section, after two or three transmissions of the same block (depending on the MCS used), the coding rate achieved is that of the convolutional code (i.e., 1/3). The segmentation mechanism provides no gain compared with the coding rate achieved due to IR. However, as IR is not mandatory on the network side, this mechanism can be useful when it is not implemented.

For the downlink direction the mechanism can be used by the BTS whenever it wants. This can be the case when owing to memory size limitation in the mobile it is not able to perform IR any more, in which case the BTS can use this mechanism to decrease the coding rate.

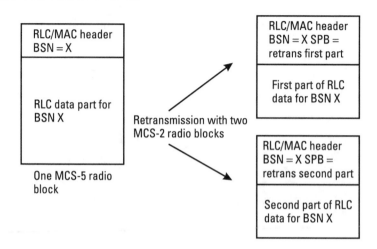

**Figure 4.6**   Example of segmentation from MCS-5 to MCS-2.

### 4.2.3.2   Choice of the MCS for Retransmission

The link adaptation mechanism is controlled by the network for both uplink and downlink transfers. The network chooses which MCS is the most suited based on the measurements performed in uplink by the BTS or in downlink by the mobile. The reaction time of the mobile to a new coding command is two block periods. The mobile shall use the new MCS two block periods after the receipt of the command.

The network orders the use of an MCS in the assignment message or in the acknowledgement message (see Chapter 5). The coding command contains any of the nine MCSs or two special commands whose meaning is explained below: MCS-5-7 or MCS-6-9. For its initial transmission, the RLC data block has to be sent with the last-ordered MCS. When MCS-5-7 (respectively MCS-6-9) is ordered, MCS-5 (respectively, MCS-6) shall be used for the first transmission and MCS-7 (respectively, MCS-9) for the following retransmission.

During an uplink transfer the mobile starts sending RLC data blocks encoded with the initial-ordered MCS. It may happen that the network does not decode some of these RLC data blocks and before their retransmission the network sends a new coding command. The mobile uses this new coding command to select the MCS that has to be chosen to retransmit the RLC data block. For the re-transmission, if the ordered MCS has not changed since the first transmission, the same one is used for the retransmission of the RLC data block. If the ordered MCS has changed, the mobile has to retransmit the RLC data block using the same or another MCS of the same family with the following rules:

- If the ordered MCS has a higher coding rate, the mobile has to use the MCS of the same family that has the nearest lowest or equal coding rate.
- If the ordered MCS has a lower coding rate, the mobile has to use the MCS of the same family that has the higher coding rate that does not exceed the ordered one.

For example, suppose that the mobile has in memory four RLC data blocks that have to be retransmitted. The first two have been initially sent using one MCS-7 radio block, and the two others have been encoded using one MCS-9 radio block, because it was the ordered MCS when the data block was segmented. If the ordered MCS is MCS-6 when the retransmission occurs, the two first RLC data blocks will be retransmitted within two MCS-5 radio blocks and the two others within two MCS-6 radio blocks.

Figure 4.7 summarizes the switching rules for the different MCSs. The box on the left indicates the previous MCS used, and the box on the right

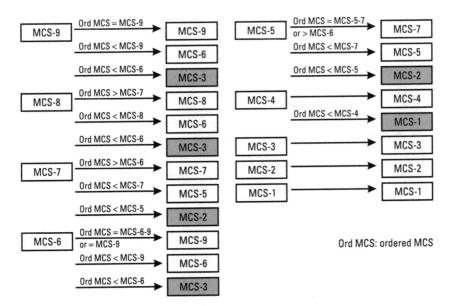

**Figure 4.7**　Switching rules between the MCSs.

the one that has to be used depending on the ordered MCS. The gray boxes are valid only when the resegmentation is allowed on the mobile side.

The figure can be read in the following way. If the initial coding of the RLC data block is MCS-9 and the last ordered MCS is the same, the block is retransmitted using the same MCS. If the ordered MCS is lower than MCS-9, the block has to be retransmitted using MCS-6. However, if the resegmentation is allowed in uplink and the ordered MCS is lower than MCS-6, the block has to be resent in two radio blocks encoded with MCS-3.

From the BTS point of view, as it manages itself the choice of the MCS there is no rule defined except that the BTS shall use an MCS from the same family to retransmit a block.

### 4.2.3.3 Choice of Puncturing Scheme for Retransmission

It was explained in Section 4.2.2.2 that the RLC data block shall first be punctured with the first scheme P1 of the commanded MCS. At the next transmission of the block, the next puncturing scheme of the commanded MCS is used and so on.

However, as described in the previous section, it may happen that one RLC data block is transmitted with one MCS the first time and, because of a change in the ordered coding command, it is retransmitted using another MCS. In this case, the receiver shall be able to perform joint decoding between the two transmissions even if the MCS used was not the same at

**Table 4.8**
RLC Data Blocks Retransmitted in New MCS

| MCS Switched from | MCS Switched to | PS of Last Transmission Before MCS Switch | PS of First Transmission After MCS Switch |
|---|---|---|---|
| MCS-9 | MCS-6 | P1 or P3 | P1 |
|  |  | P2 | P2 |
| MCS-6 | MCS-9 | P1 | P3 |
|  |  | P2 | P2 |
| MCS-7 | MCS-5 | any | P1 |
| MCS-5 | MCS-7 | any | P2 |
| all other combinations |  | any | P1 |

*From:* [2].

each transmission. In particular this is the case for MCS-6 and MCS-9, and MCS-5 and MCS-7. For example, the first time the block is transmitted with MCS-9, and for the retransmission MCS-6 is used. In this particular case, the receiver has to perform joint decoding (i.e., IR) between the two transmissions.

However for these particular cases, the management of the puncturing scheme was precisely defined in order to increase the probability of decoding. Strict rules for the management of the puncturing schemes have been defined. Table 4.8 summarizes these rules. For example, if the last transmission of one RLC data block was performed using MCS-9 and puncturing scheme P2 and the new ordered coding command is MCS-6, the block will be retransmitted using MCS-6 and P2.

## 4.3 Case Studies

### 4.3.1 IR Mechanism in Downlink

The support of IR is mandatory in the mobile. However the performance of this feature is directly linked to the memory available in the mobile for this purpose. Therefore, it has been decided as part of the EDGE standardization to define a requirement that ensures a minimum IR performance in the mobile. The goal of this case study is to explain the test conditions used to validate this requirement and to analyze the different parameters that influence the memory dimensioning. A good understanding of the test conditions is needed in order to correctly dimension the memory size. We will discuss later what the influence of the different parameters on the memory size is.

#### 4.3.1.1 IR Test

The test for IR consists of sending RLC data blocks to the mobile with MCS-9 and in measuring the achieved throughput on each time slot. The time slot allocation is equal to the maximum allocation supported by the mobile in EGPRS mode. The propagation conditions are defined by a static environment (no multipath fading profile) with an input level of –97 dBm. The required throughput that shall be achieved by the mobile is 20 Kbps per time slot. It is measured between the RLC/MAC and LLC layers.

In addition to these radio test conditions, RLC parameters are also defined. The RLC window (see Chapter 5) is equal to the highest corresponding to the multislot class of the mobile. The polling period is equal to

32 RLC blocks. This period corresponds to the number of RLC data blocks that are transmitted between two requests of acknowledgment messages by the transmitter (see Chapter 5).

The round-trip delay is equal to 120 ms. Its influence in the test case will be explained below. It has a direct impact on the performance as it defines the time that elapses between the request of an acknowledgment message sent by the transmitter and the reception of the answer at the tester side. The tester will start the retransmission of the unacknowledged RLC data blocks once the answer from the mobile has been received. Figures 4.8 and 4.9 show two examples of polling procedures that take into account the test requirements for, respectively, a single-slot mobile and a class 12 mobile (that can therefore support up to four RX time slots).

The tester starts sending RLC data blocks in sequence. In the thirty-second RLC data block (that is contained in the sixteenth radio blocks as two RLC data blocks are used per MCS-9 radio block in the test), the tester requests the sending of a downlink acknowledgement message. Due to the round-trip delay, the reception of the message on the tester side occurs $n$ radio blocks after its request (or $2n$ RLC data blocks).

The parameter $n$ is directly linked to the round-trip delay and to the number of RX slots that is supported by the mobile. As shown in Figures 4.8 and 4.9, the number of received radio blocks is equal for a single-slot mobile to 120 ms/20 ms = 6 or 12 RLC data blocks, and for a class 12 mobile with four RX time slots to 48 RLC data blocks (i.e., four times higher).

When the tester receives the acknowledgement message from the mobile, it starts to retransmit the RLC data blocks that have not been acknowledged by the mobile. For the monoslot mobile, this will start after the first forty-fourth of the sent RLC data blocks; for the class 12 mobile

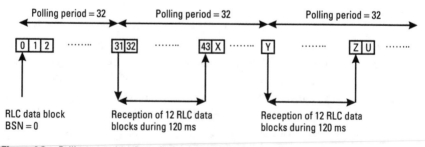

**Figure 4.8**  Polling procedure for a single-slot mobile.

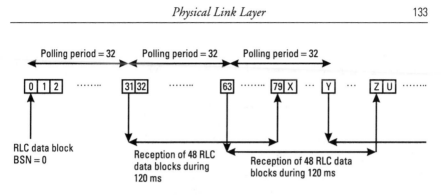

**Figure 4.9** Polling procedure for a class 12 mobile with four downlink slots allocated.

after the eightieth sent RLC data block. Thus, in Figure 4.9, X represents the BSN of the first retransmitted data block if any. Therefore, depending on the multislot class of the mobile, the impact on the memory needed for IR is not the same. The memory needed for IR increases with the number of time slots supported.

### 4.3.1.2 Influence of Different Parameters on the IR Memory Dimensioning

However, the great problem for the dimensioning of IR is linked to the following constraints:

- First, the performance of the 8-PSK equalizer/decoder. In fact the higher the performance of the 8-PSK equalizer/decoder, the lower the memory needed for IR. In fact when the equalizer performances increase, the probability of being able to decode an RLC data block will be higher; that reduces the amount of memory dedicated to IR. So it is necessary to have a good equalizer/decoder.

- Second, the performances of the equalizer/decoder improve with the number of bits used to represent the soft values (refer to Chapter 3). The smaller this number, the worse the performance. On the other hand, the amount of memory needed increases with the soft values number of bits.

### 4.3.1.3 Conclusion

Therefore, the soft values quantization has to be adjusted to find a compromise between memory size and equalizer/decoder performance in order to dimension the memory size for IR. However, the complexity of the equalizer/decoder shall not be too high in order to avoid taking all the processing power and thereby reducing the number of supported time slots for the mobile.

In order to reduce memory use, one possibility would be to increase the quantization step of the soft values after equalization. It means that the soft values used for equalization are coded on $x$ bits and when stored in memory are then coded on $y$ bits with $y < x$. This will not impact the performance of the equalizer but only that of the decoder. This is significant in case the quantization on $x$ bits is used for the first decoding and the quantization on $y$ bits is used for the next decoding.

Another important issue is the management of the soft values of the different blocks within the memory. The problem is to evaluate the number of retransmissions of the same block that have to be kept in memory by the mobile.

For example, the policy of the mobile could be to keep in memory all the soft values from the different transmissions until it decodes the RLC data blocks. Considering a block encoded with MCS-9, the receiver could store in memory the soft values corresponding to the puncturing schemes P1, P2, P3, P1, and subsequent iterations. So for the same RLC data block the amount of memory needed can be very large.

But from a performance point of view, there is surely no gain or a poor gain when the joint combining is performed on a very large number of retransmissions. So in order to manage memory as efficiently as possible, a good policy of soft values management from the different transmissions shall be implemented

In summary, the different parameters that will impact the IR dimensioning are as follows:

- The performance of the equalizer/decoder;
- The quantization of the soft values;
- The policy for the management of the soft values.

In order to optimize and dimension this memory, a good compromise between these different parameters shall be found.

### 4.3.2  Link Adaptation Implementation

This case study aims at explaining how the link adaptation can be implemented in the case of EGPRS in RLC-acknowledged mode; hence, it shows

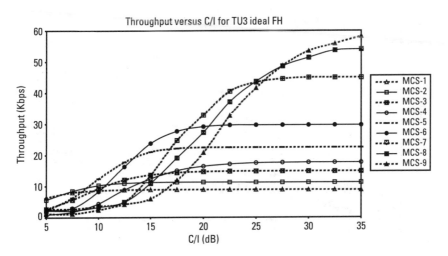

**Figure 4.10** Throughput versus C/I for the different MCSs without IR.

some of the major improvements brought about by EDGE. They are linked to the introduction of IR and the new measurement metric (BEP).

The goal of this function is the same as for GPRS; choosing the MCS that brings the highest throughput depending on the radio conditions. Figure 4.10 shows the throughput provided by the different MCSs versus the C/ I for TU3 with ideal FH.

Figure 4.10 does not take into account the IR mechanism. In order to evaluate the gain of this mechanism, new simulations have been run. These simulations consist, for a given C/I, of sending radio blocks encoded with MCS-9. The block is resent until the receiver that applied IR decodes it successfully. The achieved throughput is then computed.

Figure 4.11 shows the throughput versus C/I for the different MCSs and for MCS-9 when the latter is combined with IR.

One can observe that for almost all the C/I values, MCS-9 combined with IR achieves the highest throughput. Consequently, it is better to use MCS-9 combined with IR all the time rather than switching between the different MCSs if IR is not used. For example with a C/I equal to 20 dB, the throughput achieved with MCS-7 is lower than that achieved with MCS-9 + IR. The coding rate of MCS-7 is 3/4. So it seems that MCS-9 + IR can reach a coding rate closer to the minimum rate necessary to decode the block.

If the policy of the network is to use MCS-9 with IR all the time, there is no interest in requesting measurements from the mobile as they will not impact the link adaptation strategy. However, there is a gain in switching

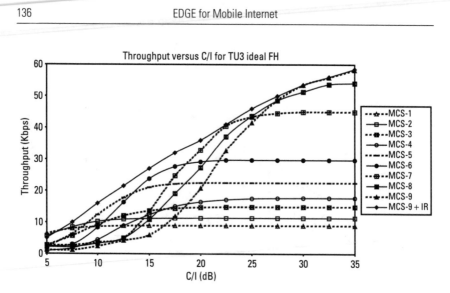

**Figure 4.11** Throughput versus C/I for all the MCSs plus MCS-9 with IR.

between the different MCSs when they are combined with the IR; estimation of link quality is really necessary. But even if the link quality estimation is not perfect, the IR mechanism will always ensure a good throughput whichever MCS was ordered. The IR allows maintaining a good throughput despite the temporary imperfections of the radio link.

As for GPRS, one of the main problems in adapting the link to the radio conditions is the estimation of its quality. The estimation of the link quality is very dependent on the environment, the use or nonuse of FH, and the velocity of the mobile.

However, with the introduction of the new measurements based on BEP, it is possible to estimate the link quality independently of the channel type. In fact, the combination of the mean BEP together with the coefficient of variation allows an accurate estimation of the channel quality. It is then possible to deduce from each combination (CV_BEP, MEAN_BEP), the corresponding "ideal" MCS to be used. One possible implementation will be to have a table giving the "ideal" MCS for each combination (MEAN_BEP, CV_BEP). Every time that new measurements are available, the network maps them onto the table and deduces the MCS that has been commanded to the MS or to the BTS.

# References

[1]   3GPP TS 05.08: Radio Subsystem Link Control (R99).

[2]   3GPP TS 04.60: Radio Link Control/Medium Access Control (RLC/MAC) Protocol (R99).

# Selected Bibliography

3GPP TS 03.64: Overall Description of the GPRS Radio Interface, Stage 2 (R99).

3GPP TS 05.01: Physical Layer on the Radio Path; General Description (R99).

3GPP TS 05.02: Multiplexing and Multiple Access on the Radio Path (R99).

3GPP TS 05.03: Channel Coding (R99).

# 5

# Impact of EGPRS on the RLC/MAC Layer

This section details the procedures that are used at the RLC/MAC layer by the mobile and the network to transfer data in EGPRS mode. It concentrates only on the procedures that are dedicated to EGPRS. For procedures that are common to GPRS and EGPRS, the reader will have to refer to Chapter 5 of [1] where the RLC/MAC procedures for GPRS are described.

As explained in Chapter 2, the introduction of EGPRS was followed by a few modifications and enhancements of basic GPRS procedures dedicated to TBF management. These differences are described in the first section of the chapter. The second section of the chapter deals with the enhancement of the RLC Protocol for the support of higher throughput thanks to the increase of the window size, a new polling mechanism, and bitmap compression for the reporting of acknowledgements.

## 5.1 New RLC/MAC Procedures Related to TBF Establishment

This section describes the impact of the introduction of EGPRS on TBF management. It concerns the establishment phase and the contention resolution phase.

### 5.1.1 Uplink TBF Establishment

The establishment of an uplink TBF in EGPRS mode is performed on PCCCH, or on CCCH when there is no PCCCH in the cell. However, because the TBF is performed in EGPRS mode, more information must be exchanged during the establishment phase than in the GPRS case. Moreover, the procedures that are used slightly differ from those for GPRS.

### 5.1.1.1   RACH/PRACH Phase

The EGPRS mobile that wants to establish a TBF in EGPRS mode requests its establishment by sending an access message. This message is sent on PRACH if it exits in the cell; otherwise it is sent on RACH. The request of a TBF in EGPRS mode can be performed only in a cell that supports EGPRS. This information is broadcast in the cell on the BCCH or PBCCH if it is present.

In order to request a TBF in EGPRS mode during the access phase, a new message dedicated to EGPRS TBF establishment was introduced. This message is the EGPRS PACKET CHANNEL REQUEST. The mobile sends this message on the RACH or PRACH when it wants to establish a TBF in EGPRS mode. However this message is optionally supported by the BTS. This is indicated in the broadcast information.

This message has the same format as the 11-bit PACKET CHANNEL REQUEST message and provides the same information, but new training sequences are used to send the message. In fact, in addition to the old one two new training sequences for sending access bursts have been introduced for EDGE. Each training sequence provides information on the EDGE capability of the mobile.

As a matter of fact, a mobile that is EDGE capable and supports 8-PSK in uplink and downlink will indicate this capability by sending an EGPRS PACKET CHANNEL REQUEST with a predetermined training sequence. A mobile that is EDGE capable but 8-PSK capable only in down-link will use the second predetermined training sequence. Moreover, these two new training sequences were designed in such a way that they can be used in parallel with the old training sequence so that it is possible to use them on both RACH and PRACH when supported. Thus the network is able to differentiate the EGPRS PACKET CHANNEL REQUEST, the CHANNEL REQUEST or the PACKET CHANNEL REQUEST message. Because of this new message, it is possible to establish a TBF in EGPRS mode in a one-phase access.

If the message is not supported by the BTS, the mobile requests the establishment of the TBF by sending a CHANNEL REQUEST message on RACH or a PACKET CHANNEL REQUEST message on PRACH using a two-phase access. The mobile will signal its EDGE capability information in the second phase of the access procedure and the network will be able to satisfy its demand.

If the mobile wants to establish a TBF in RLC unacknowledged mode, it has to request a two-phase access. In fact, the default RLC mode requested by the one-phase access procedure is the RLC acknowledged mode. The two-phase access can be requested using the CHANNEL REQUEST, PACKET CHANNEL REQUEST, or EGPRS PACKET CHANNEL REQUEST message. Note that the establishment causes that can be requested within the EGPRS PACKET CHANNEL REQUEST message are the same as in the PACKET CHANNEL REQUEST message.

### 5.1.1.2    Uplink TBF Establishment in EGPRS Mode Using One-Phase Access

This procedure occurs only when EGPRS is supported in the cell and the network is able to detect the EGPRS PACKET CHANNEL REQUEST message.

*Uplink TBF Establishment Using One-Phase Access on CCCH*

The one-phase access procedure on CCCH is shown in Figure 5.1. The EGPRS mobile triggers this procedure by sending an EGPRS PACKET CHANNEL REQUEST message on RACH. This message carries the following information: the EGPRS capability of the mobile, the support or non-support of the 8-PSK modulation in uplink and the EGPRS multislot class of the mobile. The EGPRS multislot class can be different from the GPRS one.

Note that the network cannot assign more than one time slot in the assignment message because of a limitation in the length of the IMMEDIATE ASSIGNMENT message. It will be able to extend the *mobile allocation* (MA) once the TBF has been established by sending a PACKET UPLINK ASSIGNMENT message.

The EGPRS PACKET CHANNEL REQUEST message contains the establishment cause and the random value that is used to reduce the probability that two MSs requiring the establishment of a TBF send exactly the same message in the same RACH or PRACH occurrence.

**Note**    The network does not have to respect the one-phase access request of the mobile and may force a two-phase access procedure.

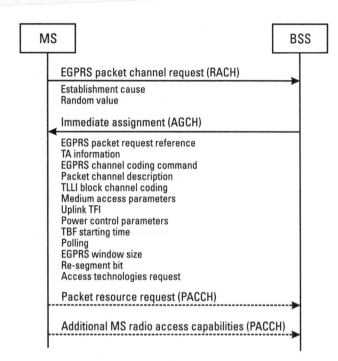

**Figure 5.1**    One-phase access establishment scenario on CCCH.

Upon reception of the EGPRS PACKET CHANNEL REQUEST message, the BSS sends an IMMEDIATE ASSIGNMENT message on AGCH that assigns resources in EGPRS mode.

This IMMEDIATE ASSIGNMENT message contains the following information:

- *EGPRS packet request reference.* This includes the content of the EGPRS PACKET CHANNEL REQUEST message (both the random value and the establishment cause) and the *frame number* (FN) in which it was received.

- *TA parameters.* This contains the initial TA and the TA index as well as the TA time slot number if the continuous TA procedure is used by the network.

- *EGPRS channel coding command.* This parameter indicates to the mobile the MCS to use for uplink data transmission.

- *Packet channel description.* This indicates the allocated time slot number, the training sequence code, and the frequency parameters.

- *TLLI block channel coding.* This is used to indicate the MCS that has to be used for data transmission during the contention resolution phase. The value could be MCS-1 or the previous EGPRS channel coding command. The contention resolution procedure is described in Section 5.5.1.6 of [1].

- *Medium access parameters.* This gives the USF value in case of dynamic allocation and the fixed allocated bitmap in case of fixed allocation.

- *Uplink temporary flow identity* (TFI). This parameter identifies the uplink TBF.

- *Power control parameters.* This indicates the downlink power control mode and the uplink power control parameters.

- *Medium access parameters.* This gives the USF value in case of dynamic allocation and the fixed allocated bitmap in case of fixed allocation.

- *EGPRS window size.* This indicates the size of the window that has to be used by the RLC Protocol.

- *Resegment bit.* This indicates whether or not the resegmentation is to be used by the mobile.

- *Access technologies request.* The network can request the MS characteristics on different frequency bands that it could possibly support. The network indicates, within the access technologies request, the frequency bands on which it requests information from the MS. If this piece of information is included within the PACKET UPLINK ASSIGNMENT message, the mobile replies by sending a PACKET RESOURCE REQUEST message and, if needed, an ADDITIONAL MS RADIO ACCESS CAPABILITIES message if the information that is requested does not fit into one message. These two messages are sent at the beginning of the TBF. This information can be used by the BSS for mobile mobility management within the network, especially when it wants to spread the mobiles within the different frequency bands in order to perform load sharing.

***Note*** If the network is unable to provide EGPRS resources to the mobile because of a congestion situation, it can allocate GPRS resources instead, if they are available. In fact, it may happen that the transceivers that are located within a cell are not all EDGE capable. So when there are no more resources on these EDGE-capable transceivers, the network may assign resources available on the other ones that are only GPRS capable.

At the end of the assignment procedure, the mobile enters the contention resolution phase. For more detailed information about this procedure the reader may refer to [1].

*Uplink TBF Establishment Using One-Phase Access on PCCCH*

Figure 5.2 describes the scenario for a one-phase access uplink establishment on PCCCH in EGPRS mode.

In order to establish a TBF in EGPRS mode using a one-phase access procedure, the mobile sends an EGPRS PACKET CHANNEL REQUEST message on PRACH indicating one of the following access types: one-phase access, short access, page response, cell update, or MM procedure.

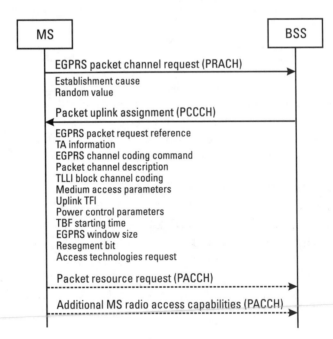

**Figure 5.2**    One-phase access establishment scenario on PCCCH.

The network sends the PACKET UPLINK ASSIGNMENT message on the PCCCH on which the request has been received. This message contains the same parameters as described in the previous section except that the medium access parameters can contain more than one time slot allocation. The medium access parameters consist of the following:

- USF values on the different allocated time slots in case of dynamic allocation or extended dynamic allocation;
- Time slot allocation, along with downlink control time slot and fixed allocation bitmap parameters in case of fixed allocation.

Note that the number of allocated slots is based on the EGPRS multislot class of the mobile. If the network is unable to allocate EGPRS resources, then it allocates GPRS resources. However, as the network does not know the GPRS multislot class of the mobile it will allocate time slots based on the EGPRS multislot class that could be nonoptimized compared to the actual one. When the mobile receives the PACKET UPLINK ASSIGNMENT message, it enters packet transfer mode and the contention resolution phase starts.

### 5.1.1.3 Uplink TBF Establishment in EGPRS Mode Using Two-Phase Access

When the EGPRS PACKET CHANNEL REQUEST message is not supported by the BTS, the two-phase access procedure shall be used to establish a TBF in EGPRS mode. When the EGPRS PACKET CHANNEL REQUEST message is supported, the two-phase access procedure is used by the mobile to establish a TBF in RLC unacknowledged mode. Depending on whether the EGPRS PACKET CHANNEL REQUEST message is supported or not, the procedures are different.

*Two-Phase Access Procedure When EGPRS PACKET CHANNEL REQUEST Is Not Supported*

The mobile requests this procedure when it wants to establish a TBF in EGPRS RLC acknowledged or unacknowledged mode and when the EGPRS PACKET CHANNEL REQUEST message is not supported in the cell.

During the first phase of the establishment procedure, the network does not know the EDGE capabilities of the mobile that requests the TBF establishment and will consider it as a GPRS mobile. So the first phase is exactly the same as the TBF establishment in GPRS mode with two-phase access (see Section 1.3.2.2 or [1] for a detailed description).

The mobile requests TBF establishment by sending a CHANNEL REQUEST message on RACH or a PACKET CHANNEL REQUEST message on PRACH. The network allocates one uplink block occurrence within the IMMEDIATE ASSIGNMENT message on CCCH or the PACKET UPLINK ASSIGNMENT on PCCCH. The mobile provides its capabilities within the PACKET RESOURCE REQUEST message that is sent in the allocated block occurrence. Upon receipt of this message the network discovers the EDGE capabilities of the mobile. It is then able to assign resources in EGPRS mode by sending a PACKET UPLINK ASSIGNMENT message that contains the same parameters as defined in Section 5.1.1.2.

*Two-Phase Access Procedure Using the EGPRS PACKET CHANNEL REQUEST Message*

The mobile requests this procedure in order to establish an EGPRS TBF in RLC unacknowledged mode. This procedure is approximately the same on CCCH and on PCCCH. Figure 5.3 describes the procedure when the establishment is performed on PCCCH.

The mobile requests the establishment of a TBF in two phases by sending an EGPRS PACKET CHANNEL REQUEST message on RACH or PRACH. The network can allocate one or two uplink resources in order to receive the capabilities of the mobile on different frequency bands. The network allocates these resources by sending an IMMEDIATE ASSIGNMENT or PACKET UPLINK ASSIGNMENT message. The network indicates within these messages the frequency bands on which it requests the mobile capabilities. In the first allocated uplink block occurrence, the mobile sends a PACKET RESOURCE REQUEST message. If a second block occurrence is scheduled, the mobile sends an ADDITIONAL MS RADIO ACCESS CAPABILITIES message.

The PACKET RESOURCE REQUEST message contains the following information:

- *TLLI.* This parameter identifies the mobile.
- *Access type.* This indicates the MS reason for requesting the access.
- *Channel request description.* This indicates the peak throughput class for the PDP context of the LLC PDU, the radio priority, the RLC mode, the type of the first LLC PDU, and the number of RLC data octets of the requested TBF.
- *MS radio access capability.* This indicates mobile capabilities in terms of multislot class, and RF power. These indications are reported for

the different frequency bands that are supported by the mobile. If this information does not fit within the PACKET RESOURCE REQUEST message, the rest is sent within the ADDITIONAL MS RADIO ACCESS CAPABILITIES message if two uplink resources have been allocated.

The network assigns the EGPRS resources by sending a PACKET UPLINK ASSIGNMENT message that contains the same parameters as described in Section 5.1.1.2.

**MS**

**BSS**

EGPRS packet channel request (PRACH)

Establishment cause
Random value

Packet uplink assignment (PCCCH)

Packet request reference
Packet channel description
Power control parameters
TBF starting time
Packet timing advance parameters
Time slot number

Packet resource request (PACCH)

Access type
TLLI
Channel request description
MS radio access capability

Additional MS radio access capabilities (PACCH)

Packet uplink assignment (PACCH)

TLLI
TA information
EGPRS channel coding command
Packet channel description
TLLI block channel coding
Medium access parameters
Uplink TFI
Power control parameters
TBF starting time
EGPRS window size
Resegment bit
ARAC retransmission request

**Figure 5.3**    Two-phase access establishment scenario on PCCCH.

***Note***   If the network assigns two uplink blocks in the first phase of the establishment and the ADDITIONAL MS RADIO ACCESS CAPABILITIES message is not decoded by the network because of bad radio conditions, it can request its retransmission by sending a PACKET UPLINK ASSIGNMENT. This request is indicated by the *additional MS radio access capabilities* (ARAC) retransmission request flag.

#### 5.1.1.4   Establishment of an Uplink TBF During Downlink TBF

This procedure, which consists of requesting the establishment of an uplink TBF within the PACKET DOWNLINK ACKNOWLEDGEMENT message during a downlink transfer is almost the same as for GPRS. The only difference concerns the sending of dedicated EGPRS parameters within the assignment message. These parameters are the EGPRS channel coding command, the EGPRS window size and the resegment indication.

#### 5.1.1.5   Special Requirements During Contention Resolution Phase

When the mobile starts its transmission after a TBF establishment in one phase, it includes its TLLI within each RLC data block until the contention resolution ends. For a detailed description of the contention resolution procedure, the reader may refer to [1]. If the EDGE mobile starts its transmission with one MCS that includes two RLC data blocks per radio block, it has to include its TLLI within the two RLC data blocks. This is needed since the two data blocks that belong to the same radio block can be independently decoded.

### 5.1.2   Downlink TBF Establishment

For the establishment of a downlink TBF in EGPRS mode, the procedure is the same as for GPRS except that the IMMEDIATE ASSIGNMENT message in case of establishment on CCCH or the PACKET DOWNLINK ASSIGNMENT in case of establishment on PCCCH contains additional parameters that are dedicated to the TBF in EGPRS mode (see Section 1.3.2.2). These parameters are as follows:

- The EGPRS window size, which gives the window size used by the RLC Protocol for the transfer of RLC data blocks in the downlink direction;
- The link quality measurement mode, which indicates whether the mobile shall report quality measurements on a time slot basis or on a

TBF basis and whether interference measurements have to be reported as well.

## 5.2 Transmission of RLC Data Blocks

The transfer of RLC data blocks in EGPRS mode reuses exactly the same concepts as in GPRS. The RLC data blocks are sent in sequence and the control is performed thanks to a sliding window mechanism. The RLC Protocol can operate in acknowledged or unacknowledged mode. When operating in RLC acknowledged mode, the acknowledgements are contained within the PACKET UPLINK ACK/NACK message in case of uplink transfer and in the EGPRS PACKET DOWNLINK ACK/NACK message for a downlink transfer.

The only difference lies in the fact that two RLC data blocks can be transmitted within one single radio block every 20 ms. This potentially leads to twice as many RLC data blocks transmissions as in GPRS. In order to cope with the higher probability that the RLC Protocol stalls, some parts of the RLC Protocol have been enhanced.

The first improvement concerns the RLC window size. However, its modification has required some changes in the acknowledgment reporting mechanism. These changes concern the way the reporting bitmap is handled as well as the polling mechanism.

### 5.2.1 RLC Window Length

As described previously, in order to enhance the RLC Protocol and to be able to support a potentially twice as high RLC data block throughput, a mechanism providing more flexibility in the management of the RLC window has been introduced. This mechanism allows the use of a variable RLC window size.

The network chooses the size of the window during the establishment of the TBF. Its range is between 64 and 1,024 in increments of 64. However, the RLC window size is dependent on the number of time slots that are allocated to the mobile. A maximum window size has been defined for each number of allocated time slots. For a given number of time slots, the window size shall be lower than this predefined maximum size. The mobile has to be able to support the maximum size corresponding to its multislot capability. Table 5.1 gives the maximum RLC window size depending on the number

**Table 5.1**
RLC Window Size Range Depending on the Number of Allocated Time Slots

| | Number of Allocated Time Slots | | | | | | | |
|---|---|---|---|---|---|---|---|---|
| | 1 | 2 | 3 | 4 | 5 | 6 | 7 | 8 |
| **Minimum Window Size** | 64 | 64 | 64 | 64 | 64 | 64 | 64 | 64 |
| **Maximum Window Size** | 192 | 256 | 384 | 512 | 640 | 768 | 896 | 1,024 |

of allocated time slots. Note that the minimum window size that shall be supported is always 64.

The RLC window size is set independently for an uplink or downlink TBF. It can be modified during the TBF but it cannot be decreased. In fact if the window size could be reduced, it could happen that some information is lost about some RLC data block states within the window, and a reliable operation of the RLC Protocol is in this case not ensured. For example, if the maximum window size is 128 and it could be reduced to 64, the state of the 64 first RLC data blocks within the window would be lost. The RLC data blocks are numbered modulo 2,048.

**Note**  The window size has a direct impact on the maximum number of RLC data blocks that can be stored in the mobile or in the network's internal memory. The larger the window size, the higher the memory needed for RLC data block management.

### 5.2.2   Compression of Acknowledgment Bitmap

When operating in RLC acknowledged mode, the receiver reports to the transmitter which RLC data blocks have been successfully received and which have to be retransmitted. The receiver uses bitmaps (series of 0,1) to indicate the RLC data blocks that have been correctly decoded.

However, as the RLC window size has been increased and can be set up to 1,024, it could be that the length of the reported bitmap reaches this size. Since the signaling is sent using CS-1, it is not possible to send more than 23 bytes of information within these blocks. Therefore, it is not possible in that case to report a complete window state bitmap within one radio block.

It has been proposed to use a compression scheme in order to put into the CS-1 encoded message as much acknowledgment information as possible. This compression mechanism is used in both the uplink and downlink direction. The compression scheme that is used is almost the same as for ITU-T T.4. The compression mechanism is very simple and consists of replacing a series of 0s or 1s by a predetermined code word representing its length. Since series of 0s and 1s are always alternating, all the receiver needs to know to recover the message is whether the initial series is made of 0s or of 1s.

For example, suppose that the receiver has to code the following acknowledgment bitmap:

00000000000000000111111111101111111111111111111111110000
11111111111111111111111111

The code words correspond to the following series:

17 times "0" is 0000011000,

9 times "1" is 10100,

1 time "0" is 0000110111,

24 times "1" is 0101000,

4 times "0" is 011,

27 times "1" is 0100100.

A table that can be found in [2] gives the mapping between the code words and a series of "0" or "1."

This leads to the following encoded bitmap:

00000110001010000001101110101000011010010 0

The transmitter has to indicate whether the first series is composed of zeros or ones. This is indicated in the acknowledgment message by the compressed bitmap starting color code. The compression is used only when the number of bits available for the bitmap reporting within the ACK/NACK message is lower than the actual window size and when there is a compression gain.

### 5.2.3 Extended Polling Mechanism for Downlink Acknowledgment Reports

In spite of the introduction of the compression mechanism for bitmap reporting, it may happen that it is not possible to report the full window state within one acknowledgment message. Then the receiver is unable to give the status of all the RLC data blocks that have been received within the window.

To solve this problem, the polling mechanism was changed for EGPRS. The main principle is quite the same as for GPRS. Whenever needed during a downlink TBF, the network polls the mobile within one radio block, to request for the sending of a downlink acknowledgment message. This message is transmitted by the receiver after $N$ frames, where $N$ is deduced from the RRBP field within the RLC/MAC header.

The difference lies in the fact that the network is able to indicate the part of the window from which it wants some information. The number of bits available for bitmap reporting is also dependent on whether or not the network requests the report of downlink measurements from the mobile. A lower rate of measurement reporting can be used in EGPRS because of the use of IR. In fact, as described in Chapter 4, IR does not require a very precise estimation of the link quality to operate reliably.

The network can request the following information using the polling request thanks to the ES/P field of the RLC header:

- The sending of a PACKET DOWNLINK ACK/NACK message without any measurement report and whose reporting bitmap starts from the oldest unacknowledged BSN (i.e., the beginning of the window). This part of the bitmap is called in the standard the *first partial bitmap* (FPB).

- The sending of a PACKET DOWNLINK ACK/NACK message without any measurement report and whose reporting bitmap starts from the highest BSN that has been reported in the previous acknowledgment message. This part of the bitmap is called in the standard the *next partial bitmap* (NPB).

- The sending of a PACKET DOWNLINK ACK/NACK message including a measurement report and whose reporting bitmap starts from the highest BSN that has been reported in the previous acknowledgement message.

Thanks to this mechanism, the network is able to acquire the state of all the RLC data blocks within the full window. Figure 5.4 shows an example of how the reporting could be managed using the extended polling mechanism.

In this example, the network requires the sending of a report starting from the beginning of the window (FPB). The mobile reports a bitmap that

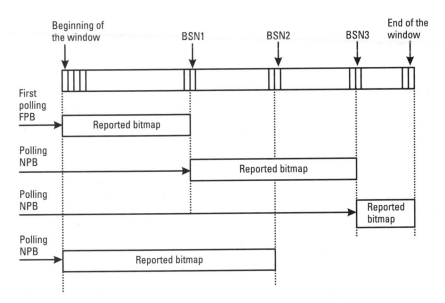

**Figure 5.4**    Polling mechanism for EGPRS.

ends at BSN1 − 1. At the second polling the network requests the sending of the next partial bitmap. The mobile reports a bitmap that starts at BSN1 and ends at BSN3 − 1. The third polling also requests the sending of the NPB. The NPB starts at BSN3 and includes the end of the window. Whatever the next polling request, the reported bitmap will start at the beginning of the window since its end has been reached in the previous one.

## 5.3    Case Study: GPRS and EGPRS Mobile Multiplexing

As described in Chapter 4, one of the main requirements when the EGPRS system was defined was the capability to multiplex GPRS and EGPRS users onto the same physical channel. This constraint mainly impacts the uplink multiplexing schemes. In the downlink direction, there is no problem since a GPRS mobile will only decode the GPRS CSs and an EGPRS mobile is able to decode both GPRS and EGPRS CSs. When a radio block encoded with one EGPRS MCS is sent in the downlink direction, a GPRS-only mobile will not decode anything.

The multiplexing problem relies on the uplink side when using dynamic allocation. With fixed allocation there is no problem in multiplexing GPRS and EGPRS users, as the allocation bitmap is sent within dedicated signaling messages addressed to the concerned mobile. But in dynamic allocation, the allocated uplink resources are assigned within the previous downlink radio block occurrence thanks to the USF. Therefore, if the network addresses a downlink radio block to an EGPRS user using 8-PSK, it will not be able to assign the corresponding uplink resource with the USF to a GPRS mobile. The USF multiplexing with a granularity of one requires that every downlink block be sent using GMSK when a GPRS mobile has to transmit on the uplink.

The network has to correlate the uplink and downlink resources allocation mechanisms so that it can allocate resources to the uplink GPRS user. If the addressed mobile for uplink allocation is an EGPRS user, there is no constraint on the downlink radio block. However, if the uplink occurrence is given to a GPRS user, the downlink block conveying this information shall be GMSK modulated.

This rule applies, for example, when two TBFs belonging to two different mobiles are in opposite directions on one PDCH. The downlink TBF is in EGPRS mode, and the uplink one is in GPRS mode. In this case, there is no EGPRS uplink transfer; thus, all the uplink resources are allocated to the GPRS user. As the uplink resources have to be assigned to the GPRS mobile using GMSK, it will not be possible to use 8-PSK in the downlink direction on the EGPRS TBF, and then the throughput will be lower. So the solution that consists of using the USF multiplexing with granularity 1 is performance-wise unattractive for the downlink direction. Figure 5.5 shows the modulation that is to be used by the network depending on whether a GPRS or EGPRS user has to transmit on the uplink.

Another solution would be to use USF multiplexing with a granularity of four blocks. When using this scheme a single USF addressed to a particular mobile holds for four successive uplink blocks occurrences. Consequently,

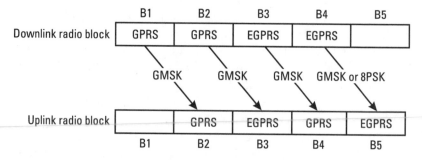

**Figure 5.5**    Multiplexing of GPRS and EGPRS users in uplink with USF granularity 1.

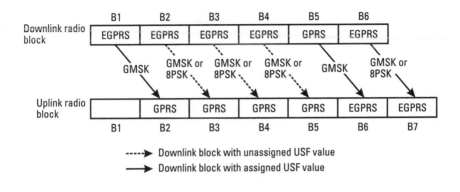

**Figure 5.6** Multiplexing of GPRS and EGPRS users in uplink with USF granularity one and four.

only every fourth downlink block has to be sent using GMSK modulation when a GPRS user has to transmit on the uplink (a GPRS user is able to detect the USF even if the radio block has been encoded using MCS-1 to MCS-4; see Chapter 4). The average fraction of radio blocks that have to be transmitted using GMSK will then be lower than one fourth, considering that both GPRS and EGPRS mobiles are multiplexed on the uplink. One possible solution could be to assign uplink TBF for GPRS mobiles with an USF granularity of four and with a USF granularity of one for EGPRS mobiles. This allows keeping more flexibility for the multiplexing of EGPRS users on the uplink. Figure 5.6 shows one example of uplink USF multiplexing with a granularity of four for the GPRS users.

Two different mechanisms are available for the dynamic multiplexing of GPRS and EGPRS users on the same physical channel. Dynamic allocation with granularity of four provides a more efficient solution (in terms of throughput on the downlink) than dynamic allocation with a granularity of one. However, this multiplexing requires the synchronization of both uplink and downlink resources schedulers in order to force the usage of GMSK on the downlink when a GPRS mobile is to use the uplink.

# References

[1] Seurre, E., P. Savelli, and P. J. Pietri, *GPRS for Mobile Internet*, Norwood, MA: Artech House, 2003.

[2] 3GPP TS 04.60 Radio Link Control/Medium Access Control (RLC/MAC) Protocol (R99).

# Selected Bibliography

3GPP TS 03.64 Overall Description of the GPRS Radio Interface, Stage 2 (R99).

3GPP TS 04.18 Radio Resource Control Protocol (R99).

3GPP TS 05.01 Physical Layer on the Radio Path; General Description (R99).

3GPP TS 05.02 Multiplexing and Multiple Access on the Radio Path (R99).

# 6

# Wireless Application Protocol

## 6.1   General Interest of Wireless Application Protocol (WAP)

WAP is at the convergence of several evolving network technologies: wireless data, telephony, and the Internet. These technologies were developed quickly and independently long before thought to combine them and exploit them for new services.

The integration of these technologies in a wireless network requires taking into account the MS constraints and the limitations of the wireless data network. On one hand, the MS presents an environment less adapted than a fixed terminal because of its screen size, less powerful CPU, less memory for program code and data, and limited autonomy. On the other hand, the wireless network has shown its limitations compared with the wired network in terms of less bandwidth provided to the users, increased latency time, more than connection setup failure, and less connection stability.

The WAP specifications define a complete environment (communication protocol, language for application development, application environment) by taking into account the constraints of wireless handheld devices and of the wireless network. The use of memory and CPU and power consumption shall be optimized in the MS while the narrowband bearers with high latency times shall be reduced in wireless network.

The WAP Forum is in charge of the WAP specifications. When it was established, some high-level requirements were defined in order to satisfy the operator needs. The requirements are interoperability, scalability, efficiency, reliability, and security. Interoperability means that any MS stamped with WAP technology must operate with any mobile network supporting WAP services. Scalability allows the personalization of services according to user needs. Efficiency allows ensuring a certain QoS related to the behavior and the characteristics of the mobile network. Reliability allows providing a consistent platform on which the services are based.

## 6.2  WAP Forum

At the origin, the WAP standard was created by a software company, Unwired Planet, which had built browser solutions for MSs. Unwired Planet, renamed Phone.com and then OpenWave, succeeded in originating the WAP Forum in 1997 with the main mobile manufacturers in order to define a set of WAP specifications. Now around 450 members have joined the WAP forum.

The WAP specifications are structured in the WAP release. A new WAP release is published every year. The WAP releases contain the specifications listed here:

- WAP 1.1—published in June 1999.
  - Protocol stack (Wireless Session Protocol, Wireless Transaction Protocol, Wireless Datagram Protocol);
  - *Wireless application environment* (WAE) (Wireless Markup Language, Wireless Markup Language Script);
  - Security (wireless transport layer security).
- WAP 1.2—published in June 2000.
  - Push services (service indication, service loading);
  - Content adaptation according to terminal and network configuration (user agent profile);
  - *Wireless telephony application* (WTA);
  - *Wireless identity module* (WIM).
- WAP 2.0—published in August 2001.
  - Browsing on XHTML site (wireless profiled HTTP, wireless profiled TCP, WML2);
  - WAP provisioning;

- *Multimedia Messaging Service* (MMS);
- Use of pictogram;
- End-to-end security (transport layer security);
- WAP certificate management (WAP public key infrastructure);
- Data synchronization (WAP SyncML);
- Use of external peripherals (external functionality interface).

## 6.3 WAP Services

### 6.3.1 Browser Services

The WAP browser is based on the WAE specifications. The WAE forms a common application(s) environment for operators and service providers in order to create applications and services on a large variety of wireless platforms. Since the WAE follows the *World Wide Web* (WWW) model, all contents handled by the browser are specified in similar formats as the existing Internet standard formats but in an optimized way.

The WAE is able to manage the following formats:

- *Wireless Markup Language* (WML)—a markup language built from *Extensible Markup Language* (XML), it is used to define the message content and a user interface for the limited capabilities of a handheld terminal.
- *WAP Binary Extensible Markup Language* (WBXML)—a compact binary representation of the XML language, used to reduce the length of XML documents.
- *Wireless Markup Language Script* (WMLScript)—a script language based on a subset of JavaScript providing script facilities; it is used conjointly with WML to increase the browser efficiency by providing additional capabilities not supported by WML.
- WML2—an evolution of WML for a convergence with *Extensible Hypertext Markup Language* (XHTML), which is a reformulation of *Hypertext Markup Language* (HTML).
- Cookie—a small chain of character to identify a user context.
- *Cascading Style Sheet* (CSS)—a style sheet to control the presentation of XHTML or WML2 document.

- vCard—a phonebook data exchange format, that is defined by the Versit consortium and managed by *Internet Mail Consortium* (IMC).

- vCalendar—a calendar data exchange format, defined by the Versit consortium and managed by IMC.

- *Wireless BitMaP* (WBMP)—a graphic format for several compressed image format types that is independent of the handheld terminal form factor.

### 6.3.1.1  WML 1.3

WML provides means for the general presentation of texts and images to the user by proposing a set of various markup elements such as emphasis elements (e.g., bold or italic), or line-break models (e.g., line wrapping).

WML 1.3 is a tag-based document language and is specified as an XML document type. It was inspired by HTML4 and by *Handheld Device Markup Language* (HDML2). It was introduced in the WAP 1.1 release. WML is defined for small narrowband device constraints such as small display, limited user-input facilities, narrowband connections, limited memory resources, and limited CPU.

The WML data are structured in cards and in decks. Cards specify one or more units of user interface elements such as selection menu, text screen, among others. A deck is a group of cards identified by a *Uniform Resource Locator* (URL) and elementary units for WML content transmission similar to an HTML page. The user browses through a set of WML cards, reviews the content of a WML card, navigates to another card, and fetches a new deck on origin servers.

### 6.3.1.2  WMLScript

WMLScript, based on JavaScript, is adapted to narrowband devices. The script improves the browsing capabilities by providing new functions that are not supported by WML. It includes advanced user interface functions such as the validity check of user input or message display, mathematical functions, and a means to access the device and its peripherals.

These capabilities supported only by WMLScript make it possible to increase the browser efficiency by limiting the number of round trips to the origin server involving a reduction of the useful bandwidth for data exchange between the server and the client.

### 6.3.1.3 WML2

WML2 is an evolution of WML in the WAP 2.0 release for a convergence with XHTML. XHTML is the new presentation language standard defined by the *World Wide Web Consortium* (W3C) which should replace HTML4. A semantic backward compatibility is ensured in the WML2 specifications so that a browser compliant with WML2 can operate with a WML 1.x document. A conveyance with existing and evolving Internet standards has been taken into account by the WAP Forum in order to take advantage of these existing standards, especially the ones from the W3C—XHTML basic, CSS Mobile Profile.

Access to device resources has been optimized for an efficient use due to the relatively limited memory and the CPU power, as well as the small form factor such as the limited display area and the restricted input facilities.

WML2 specifies the user interface elements in an abstract manner. This approach provides maximum flexibility and enables creation of distinct user interfaces.

### 6.3.1.4 Cookie

The cookie mechanism was introduced with the support of HTTP1.1 in the WAP 2.0 release for session context management. The cookie allows specifying a user context (session number for multipage transactions, various preferences, temporary variables). The cookies may be saved in the terminals like in the WWW environment, but the WAP 2.0 release adds the ability to save cookies in an intermediary proxy. The storage of cookies in a proxy enables saving the storage memory in the handheld terminal and also enables users to retrieve their context if they changed their handheld terminals.

### 6.3.1.5 CSS

The CSS defined in the WAP specifications contains a subset of CSS2 specifications from the W3C- and WAP-specific extensions. The WAP CSSs are used to control the presentation of an XHTML document by separating the graphic enrichment of a document from its content. The text/css *Multipurpose Internet Mail Extension* (MIME) media type represents the CSS defined in the WAP specifications.

The WAP CSSs use many core CSS functions defined in the CSS2 specifications (e.g., selector, syntax, parsing, data types, cascading, inheritance, and media types). Moreover, the WAP CSSs reuse properties defined in the CSS2 specifications such as margin, padding, border width, border color, border style, border shorthand, foreground color, background color, background images, font family, font style, font variant, font weight, font size, shorthand font, lists, texts (indentation, alignment, decoration, transformation, white space, visual effects), and visual formatting (display, float positioning, float flow control, content width and height).

The WAP specific extensions for CSS WAP are as follows:

- Marquee—defines simple animations of text such as the scrolling function.
- Access keys—defines a way to activate an element using a keypad key or key combination.
- Input—defines the type of data that the input element accepts.

### 6.3.2  Push Services

The push service allows transmission of information toward MSs in the client/server model without explicit request from the client. A push operation is always initiated by a *Push Initiator* (PI), an application running on a Web server. The push service allows to deliver various media types between a PI and a client (browser application on a handheld terminal) as listed here:

- MIME;
- Provisioning;
- *Wireless Telephony Application* (WTA);
- Multimedia messaging services;
- *Service indicator* (SI);
- *Service loading* (SL);
- *Cache operation* (CA).

#### 6.3.2.1  SI

The SI media type allows description of a short message with a *Uniform Resource Identifier* (URI) indicating a service to be consulted. The user can either start the service indicated by the URI upon receipt of an SI message or postpone the service for later handling. The reading of e-mails is an example of SI services: E-mail arrival is notified by a push message to the user, and the user decides either to read its content immediately or to defer the reading.

The PI may improve the behavior of the browser by providing the following mechanisms in the SI media type:

- *User-intrusiveness levels.* Different levels of priority are assigned to affect browser behavior.
- *Deletion.* This deletes an already received SI.
- *Replacement.* This replaces an old received SI with a new one.
- *Handling of out-of-order delivery.* This discards an SI received if it is older than any similar SI already saved.
- *Expiration.* This invalidates a received SI after a certain amount of time.

### 6.3.2.2  SL

The SL media type contains a URI indicating a content to be loaded without end-user confirmation. The PI may control if the browser must immediately load the indicated service or if the browser must wait the end of its activities in progress. Further, the PI controls whether the downloaded service shall be executed or shall be placed in cache memory.

### 6.3.2.3  CO

The CO media type provides a means to invalidate specific objects or services in the cache memory of the browser. The result of this operation is that the invalidated object shall be reloaded from the origin server when the user wants to have access to it. The CO is useful for applications such as the mailbox. In fact, users usually check their mailboxes less frequently than messages arrive. The sending of a CO makes it easier to reduce the network load than sending the updated mailbox content on every change.

## 6.3.3  WTA Services

The WTA provides a WAP browser with a means to have access to simple telephony services. The WTA specifications are a set of specific extensions related to telephony to set up and receive phone calls. A special WTA library, the *WTA interface* (WTAI) public library has been defined. The functions of this library may be called from any WAE application and allow access to telephony functions.

Three WTAI libraries have been defined, as follows:

- *Network common WTA library.* This library brings together the most common functions available for all network types (e.g., setting up calls). Access to these functions is reserved for a WTA client.

- *Network-specific WTA libraries.* These libraries contain functions that are specific to one network type or to one given operator (e.g., voice mail). Access to these functions is reserved for a WTA client.

- *Public WTA library.* This library provides simple functions (e.g., call configuration) that are available for any client.

### 6.3.4  Security Services

WAP specifications fulfill the five basic security requirements—authentication, confidentiality, integrity, authorization, and nonrepudiation.

The authentication mechanism enables authentication of either the end user or the server. The identity of the end-user attempting a connection is checked by the server, while the user may verify that the addressed server is the right one. This may be achieved with a public key cryptographic mechanism, or with a certificate.

The confidentiality mechanism ensures that the exchanged messages are not understandable by someone other than the transmitter and the receiver. This is achieved by using encryption.

The integrity mechanism provides a way to check that the received message has not been changed by an intruder. A message digest is appended to the message for this purpose. It corresponds to a condensed form of the message body. If one bit of the message body is changed by the intruder, it results in a change of several bits in the message digest.

The authorization mechanism checks that the user has the right to access a given service. Certificates with a validity period are given to the user for this purpose.

The nonrepudiation mechanism certifies that a received message may not be invalidated subsequently. It gives the proof about the author and the delivery date of the message. It is based on the digital signature, which is an encrypted message digest.

The *wireless transport layer security* (WTLS) provides from the WAP 1.x release these services between the WAP client and the WAP gateway or WAP proxy. *Transport layer security* (TLS) provides the same services between the WAP proxy and the application server. The WAP 2.0 release has introduced end-to-end secure connections between the WAP client and the *Hypertext Transfer Protocol* (HTTP) servers.

The WIM enables application-level security functions. WIM provides maximum security for the storage of user certificates, server certificates, private keys, and the like. The main requirement for a WIM module is to be tamper-resistant, to prevent any attempt to extract or modify information in the module. A WIM may be included or not in a SIM card. This feature was introduced in the WAP 1.2.0 release.

The WAP 2.0 release has also introduced the *public key infrastructure* (PKI), which is intended to provide a trusted relationship needed for authentication of clients and servers by enabling the exchange of user certificates and server certificates for secure connections.

### 6.3.5 User Agent Profile

The *user agent profile* (UAProf) contains information used for content formatting purposes. This specification starts from the fact that the contents delivered to WAP devices do not take into account their ever-divergent range of input and output capabilities, network connectivity, and levels of scripting language support. The goal of the UAProf is to allow all elements of networks (proxies, support servers, application or content servers) to adapt their contents according to the WAP device type. The UAProf contains the hardware and software characteristics of the device as well as information about the network to which the device is connected. It does not contain any information about the user preference. The UAProf is also referred to as *capability and preference information* (CPI). The formalism used for CPI description is based on *composite capabilities/preference profile* (CC/PP); this high-level formalism was defined by the W3C for describing and transmitting information about the capabilities of Web clients and the display preferences of Web users.

A CPI profile is composed of the six following components, each containing a collection of attribute-value pairs or attributes:

- *Hardware Platform*. This contains the hardware characteristics of the device (screen size, color capabilities).

- *Software Platform*. This describes the operating environment of the device (operator system vendor, version, list of audio and video encoders).

- *Browser UA.* This defines the set of attributes to describe the HTML browser application (HTML version).

- *Network Characteristics.* This contains information about the network-related infrastructure and environment such as bearer information (latency, reliability).

- *WAP Characteristics.* This defines the set of WAP capabilities supported on the device (WMLScript libraries, WAP version, WML deck size).

- *Push Characteristics.* This defines the set of push specific capabilities supported by the device (supported MIME type, maximum length of received message).

### 6.3.6  Provisioning Services

The provisioning services provide a transparent and secured means to configure a WAP client with a minimum of user interaction, for the access of a WAP client to WAP services through a WAP infrastructure. The provisioning provides the WAP client with the addresses and the methods to access various elements of the WAP infrastructure.

The provisioning services specify the parameters related to the following:

- *Network access point* (NAP) between the wireless and wire-line networks, often a *remote access server* (RAS);
- Various proxies (e.g., master pull proxy, push proxy, MMS proxy);
- Server certificates shared secret for authentication;
- *Trusted provisioning server* (TPS);
- Specific content location;
- Various servers (e.g., WTA server or PKI portal).

### 6.3.7  MMS

The MMS provides messaging operations with a variety of media types. MMS allows sending and receiving messages containing a rich set of contents (text, picture, video, sound). MMS provides a service of non–real-time message delivery. It also provides the ability to interact with other messaging systems such as Internet e-mail systems based on the SMTP, POP, and IMAP transport protocols, as well as legacy wireless messaging systems (paging and SMS systems).

### 6.3.8 Synchronization Services

Data synchronization enables synchronizing replicated data, such as business data stored in several devices and handled by a *personal information manager* (PIM). A PIM is an application located in a client or in a server that manages personal information such as calendar, phone book, notes, and messages.

Data synchronization over WAP is based on the SyncML protocol published by the SyncML Initiative in December 2000. SyncML is a high-level protocol based on the XML language. It allows exchange of data objects and related information between a local and distant database in order to synchronize their contents. Data synchronization over WAP relies upon the underlying session protocol (i.e., either the WSP layer in WAP 1.x architecture or the HTTP layer in the WAP 2.0 architecture).

SyncML supports the synchronization of data format such as vCard and vCalendar. These formats were specified by the VERSIT consortium and are managed by the IMC. The vCard format enables encoding of electronic business card, while the vCalendar format enables encoding of electronic calendars and schedules.

### 6.3.9 External Functional Interface

The *external functional interface* (EFI) provides methods for a WAP application to have access to an external functionality, which is a component or an embedded application executed outside of the WAP browser. With this mechanism a WAP client may control, for example, a digital camera in order to retrieve images for another storage or for a display on the screen phone. WAE and WTA applications have access to external functionality through the *EFI Application Interface* (EFI AI).

## 6.4 WAP Architecture

### 6.4.1 Architecture Overview

WAP architecture is based on a client/server model as the Internet WWW architecture. One of WAP's significant enhancements is the integration of push mechanism. Unlike the pull mechanism, no explicit request from the client is sent to the server for the push mechanism before the server transmission. The request-response mechanism is commonly referred to as the pull mechanism.

The general WAP architecture is composed of a client, a gateway, a proxy, application servers and supporting servers. The characteristics of these are as follows:

- *Client.* A client refers to a browser located in a handheld terminal. It initiates WAP requests toward an application server in order to retrieve WAP contents.

- *Gateway.* A gateway provides protocol conversion between the WAP 1.x protocol stack (WSP, WTP, WTLS, and WDP) and the WWW protocol stack (HTTP, SSL, TCP). It may contain a content encoder and decoder to convey the WAP content in a compact format under the underlying link.

- *Proxy.* A proxy provides functions fulfilled by a gateway by translating requests from a wireless protocol stack (e.g., the WAP 1.x stack) to the WWW protocols. It may manage a UAProf that describes the client's capabilities and personal preferences. It may include content encoders and decoders to translate WAP content into a compact format. It may manage a cache of frequently accessed resources.

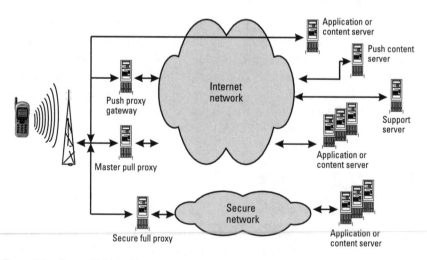

**Figure 6.1**    General WAP architecture.

- *Application server.* This is the server on which a given resource is located. It is also called the origin server.

- *Supporting servers.* These servers provide various services to devices, proxies, and application servers (e.g., provisioning servers and MMS servers).

There are two models for the architecture of the WAP 2.0 release, a model with proxy and a model without proxy. In the first model, there is a proxy residing between the client and the application server. In the second model without proxy, there is a direct HTTP link between the client and the application server. In the WAP 2.0 release, the concept of WAP gateway has disappeared in favor of a WAP proxy.

As shown in Figure 6.1, the WAP client communicates with application servers either through a number of appropriate proxies or directly.

## 6.4.2 WAP Configurations

The client communicates with the application server by using various WAP communication protocols depending on the WAP configuration.

Figure 6.2 gives the protocol stack used for the WAP 1.x release. The WAP gateway performs the protocol conversion between the WAP protocol stack (WDP, WTLS, WTP, WSP) and the IP stack (TCP, SSL, HTTP).

The WAP 2.0 specification has introduced the HTTP stack between the WAP client and the application server. The use of this stack enables a direct access to a content server on the Internet. Figures 6.3 to 6.5 show several WAP configurations in WAP 2.0.

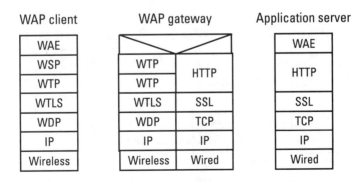

**Figure 6.2** Protocol stack using WAP gateway for WAP 1.x release.

WAP client          WAP proxy          Application server

| WAE | | | | | | WAE |
|---|---|---|---|---|---|---|
| HTTP* | | HTTP* | HTTP | | | HTTP |
| TCP* | | TCP* | TCP | | | TCP |
| IP | | IP | IP | | | IP |
| Wireless | | Wireless | Wired | | | Wired |

**Figure 6.3**   Protocol stack used by WAP HTTP proxy with profiled TCP and HTTP.

Figure 6.3 gives a WAP configuration for a WAP HTTP proxy between the wireless and the wired networks. A wireless profile of TCP (TCP*) and a wireless profile of HTTP (HTTP*) allow enhanced performance.

Figure 6.4 shows a WAP configuration with a WAP HTTP proxy that has established a connection-oriented tunnel to the application server in order to provide end-to-end security (use of the TLS Protocol).

Figure 6.5 shows a direct access between the WAP client and the application server through the Internet.

## 6.4.3   WAE

The WAE specifies all the WAP architecture elements related to application execution and application specification. On the client side, the WAE architecture is based on user agents. These latter are software implemented in a handheld terminal, which interpret contents such as WML. On the application server side, the content generator (e.g., CGI scripts) provides standard

WAP client          WAP proxy          Application server

| WAE | | | | | | WAE |
|---|---|---|---|---|---|---|
| HTTP | | | | | | HTTP |
| TLS | | | | | | TLS |
| TCP* | | TCP* | TCP | | | TCP |
| IP | | IP | IP | | | IP |
| Wireless | | Wireless | Wired | | | Wired |

**Figure 6.4**   Protocol stack used by WAP HTTP proxy for TLS tunneling.

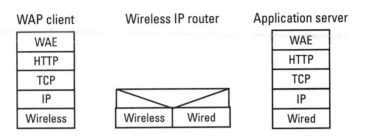

**Figure 6.5** Protocol stack used for direct access.

content format as a response of a request from a user agent located on a WAP client. The WAP gateway or WAP proxy compresses the standard content format by using the compressed WBXMP format over the mobile network. Figure 6.6 shows the general WAE architecture.

### 6.4.3.1 Naming

The WAP content is identified by a URL on an application server or origin server. WAE uses the HTTP semantic for the URL scheme. Further, WAE also provides a way to identify local resources by reserving a set of URI schemes. The URI corresponds to the URL for naming resources stored on application servers. The use of this mechanism does not imply a communication between the user agent and the application server. For example, the URI enables access to some WTAI function libraries.

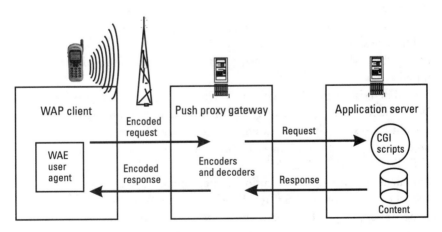

**Figure 6.6** WAE general architecture.

### 6.4.3.2    WAE Client

The WAE client supports several user agents (e.g., WML user agent, WTA user agent, MMS user agent); each user agent handles a given environment within a WAE client. It means that the user has access to a specific WAP service via an associated user agent. The WAE user agents handle various formats based on the WWW technology such as markup languages (WML/WML2), scripting language (WMLScript), graphic format (WBMP), electronic business card objects (vCard), and calendar information (vCalendar).

The main user agent handles WML decks and WMLScript. It initiates WML exchange with the origin server on user request, the service located on the origin server being identified by a URL. The WML user agent may submit an additional request for WMLScript also referenced by a URL if it encounters a WMLScript reference in a WML deck returned by the origin server. The script is compiled on the way back in order to generate a compiled bytecode that is appropriate for the WAE client. Next, the compiled bytetecode is forwarded to the WAE client to be executed.

## 6.4.4    WAP Protocol Layers

### 6.4.4.1    Wireless Session Protocol (WSP)

The WSP layer provides services suited to browser applications. Two session modes are defined: connection-oriented mode and connectionless mode. The WSP layer in connection-oriented mode operates above the *Wireless Transaction Protocol* (WTP). The WSP layer in connectionless mode operates above the secure or nonsecure datagram transport services provided by the WDP. The services provided by the WSP layer in connection-oriented mode are reliable, whereas the services provided in connectionless mode may be unreliable.

The goal of the WSP layer is to allow exchange of content between a WAP client and server. It is why the WSP layer reuses the HTTP 1.1 functionalities—especially the extensible request-reply methods, composite objects, and content-type negotiation. The method invocation defined by HTTP (GET, OPTIONS, HEAD, DELETE, TRACE, POST, PUT) allows the client to request the execution of appropriate operations by the server and the return of a result to the client. The WSP messages (request–response) exchanged between a client and a server are composed of headers (meta-information) and data field; the header structure enables qualifying the content type, character set encoding, languages, and the like; the data field is only meant for push facilities and for some invocation method facilities. Contrary to HTTP, the content exchanged between a WAP client and a server uses a compact encoding.

The WSP layer provides a set of functions to manage a WSP session lifetime (session connection, session disconnection, session suspension, session resume). The WSP layer provides push data transfer from a server to a client in opposition to pull data transfer, which is based on request response exchange between a client and a server. Three push mechanisms are defined as follows:

- *Confirmed data push within an existing session context.* The server can push data during an existing session context. After push delivery, it waits for an acknowledgment.

- *Nonconfirmed data push within an existing session context.* The server can push data during an existing session context without waiting for a confirmation of the push delivery.

- *Nonconfirmed data push without existing session context.* The server can provide a data push service without an existing session in order to send one-way messages over an unreliable transport.

The WSP layer provides the negotiation of extended capabilities between peer WSP entities, allowing an acceptable level of service. In addition, WSP provides an attachment of header information (meta-information) to the acknowledgement of a transaction. This mechanism is used to communicate specific information about the completed transaction. Simultaneous asynchronous transactions, allowing multiple requests to be made simultaneously, are optionally supported by the WSP layer.

### 6.4.4.2  WTP

The WTP layer is a transaction-oriented protocol running on top of a datagram service. The WTP layer improves the reliability over datagram services by ensuring retransmissions, duplicate removal, and acknowledgment of PDUs. Each transaction is identified by an unique *transaction identifier* (TID). The WTP layer improves the efficiency of connection-oriented services, as there is no excessive overhead on the communication link. This is due to the fact that there are no explicit connection setup or tear-down phases. The WTP layer is message-oriented and is designed for services oriented toward transactions such as browsing; WTP contains two types of messages—data messages and control messages. Data messages carry user data, whereas control messages are used for acknowledgment, and error reporting. Each WTP message is contained in the datagram payload.

Three classes of transactions are defined as follows:

- *Class 0*—unreliable invoke message with no result message. This class is related to an unreliable datagram that can be used by applications such as nonconfirmed data push; in class 0, the receiver does not acknowledge the invoke message and the initiator of the transaction does not retransmit the invoke message.

- *Class 1*—reliable invoke message with no result message. This class is related to a reliable datagram service that can be used by applications such as confirmed data push. In class 1, the receiver acknowledges the invoke message and the initiator of the transaction may re-transmit the invoke message if needed.

- *Class 2*—reliable invoke message with exactly one reliable result message. This class is related to basic invoke/response transaction service; in class 2, the receiver sends one reliable result message upon receipt of an invoke message or sends an acknowledgment indicating "hold on" before sending the result message if the receiver takes a long time to answer in order to avoid unnecessary retransmission of the invoke message.

In addition, the WTP layer provides the following functionalities:

- *Information in last acknowledgment.* This is the ability to attach information in the last acknowledgment of the transaction in order to transport a small amount of information related to the transaction. For instance this information may be performance measurements collected to evaluate the QoS perceived at the user level.

- *Retransmission until acknowledgment.* This mechanism is used to ensure a reliable data transfer in case of packet loss by acknowledging the packets that are correctly received. The incorrectly received packets are retransmitted.

- *User acknowledgment.* The WTP user may confirm each received message from the WTP server.

- *Concatenation and separation.* The concatenation mechanism makes it possible to convey multiple WTP PDUs in one datagram *service data unit* (SDU), while the separation mechanism enables extraction of several WTP PDUs from one datagram SDU.

- *Asynchronous transaction.* This mechanism enables initiation of simultaneous asynchronous transactions before receiving the response to the first transaction.

- *Transaction abort.* This mechanism is used to abort an outstanding transaction.

- *TID verification.* This is a three-way handshake procedure between an initiator and a responder used to minimize the number of transactions being replayed as a result of duplicate packets.

- *Transport information items (TPIs).* This enables support of additional information (TPIs) in the header of a WTP PDU for future extensions of the protocol.

- *Error handling.* A transaction is aborted when an unrecoverable error is detected.

- *Segmentation and reassembly.* The segmentation enables segmenting a WTP PDU in several packets if its length exceeds the *maximum transmission unit* (MTU) of one packet. Selective retransmissions may be used to request one or multiple lost packets. The reassembly mechanism enables reassembly of several packets into one WTP PDU.

### 6.4.4.3 WDP

The WDP layer is implemented in WAP for various network types. It allows adapting the upper layer protocols (WSP, WTP, WTLS) to the specific features of the underlying wireless network. The adaptation layer is specific to the underlying wireless network. By providing a common interface to the upper layer protocols, the latter are able to operate over various multiple network types.

Several protocol profiles were defined in order to map the WDP layer onto different network types with different characteristics. These profiles enable the WDP layer to operate over the digital cellular systems such as GSM (SMS, USSD, circuit-switched data, GPRS, EGPRS), DECT, PHS, TETRA, and others. Figure 6.7 gives the protocol profile for the WDP layer operating over the GPRS bearer service.

The WDP layer supports several simultaneous communication instances from a higher layer over the same underlying wireless network. The higher layer above WDP is identified by a port number. If the underlying wireless net-

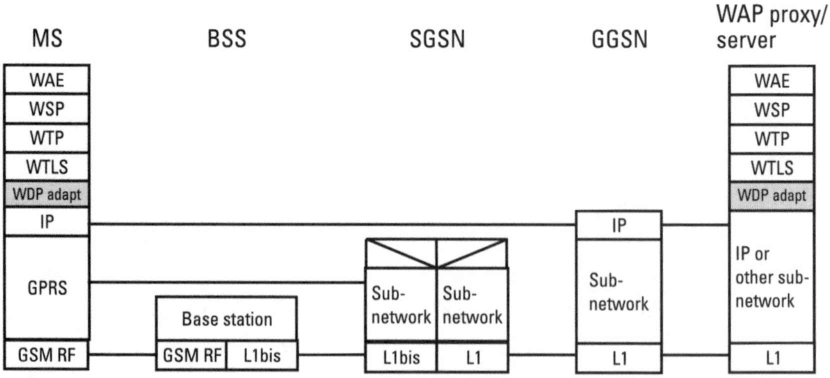

**Figure 6.7**    WDP over GSM GPRS.

work does not provide the segmentation and reassembly functions, these latter are implemented at the WDP layer in a bearer-dependent way.

### 6.4.4.4    Wireless Control Message Protocol (WCMP)

The WCMP is used by the WDP nodes and wireless data gateways in non-IP environments for diagnostic purposes. It reports errors encountered in datagrams processing at the WDP layer.

### 6.4.4.5    WTLS

The WTLS is a security layer protocol based on the TLS Protocol, previously known as *secured socket layer* (SSL). The WTLS layer runs over the WDP layer and provides secure transport over datagrams. WTLS has been optimized for use over narrowband communication channels with relatively long latency.

The WTLS Record Protocol receives uninterpreted data to be transmitted from the upper layer on the transmission side. It performs data compression (optionally), appends a *message authentication code* (MAC), encrypts data, and forwards the record to the WDP layer. The WTLS Record Protocol receives the record from WDP on the reception side, decrypts the received data, checks the MAC, performs the data decompression, and delivers data to the upper layer.

The Handshake Protocol allows negotiation between the peer WTLS entities on a secure session by negotiating the compression algorithm, the bulk data encryption algorithm, and the MAC algorithm. It is based on three messages. The client and the server exchange hello messages to establish the following attributes: protocol version, key exchange suite, cipher suite for encryption, compression method, key refresh, and sequence number mode. Following the hello message, the server sends its authentication certificate. The server may request a certificate from the client.

At this point the Change Cipher Spec Protocol is used to notify a change in the ciphering strategies. It consists of a single message sent either by the client or the server to notify the other end that subsequent records will be protected under the newly created bulk data encryption algorithm and keys.

The Alert Protocol makes it possible to inform the other WTLS peer entity that an error has occurred. The error can be fatal or not and may lead according to the severity of the message to close the secure connection. Upon detection of an error, the detecting peer sends an alert message to the other end.

Three classes exist for WTLS. Each WTLS class is defined by a set of mandatory features and by a set of optional features. A WAP client compliant to a given WTLS class implies that it supports at least all mandatory features related to this WTLS class and may support the optional features. Table 6.1 gives the definition of WTLS classes (M for mandatory, O for optional).

### 6.4.4.6  IP

In order to have access to an HTTP server, the WAP stack in the release 2.0 includes the HTTP which provides the hypermedia transfer service over

**Table 6.1**
Definition of WTLS Classes

| Feature | Class 1 | Class 2 | Class 3 |
|---|---|---|---|
| Public-key exchange | M | M | M |
| Server certificates | O | M | M |
| Client certificates | O | O | M |
| Shared-secret handshake | O | O | O |
| Compression | Not defined | O | O |
| Encryption | M | M | M |
| MAC | M | M | M |
| Smart Card Interface | Not defined | O | O |

secure and nonsecure connection-oriented transports. There are two network configurations for access to the HTTP server from the WAP client as shown in Section 6.4.2—either a direct access or via a WAP proxy. For this purpose, the WAP stack includes *Wireless Profiled HTTP* (W-HTTP), *Wireless Profiled TCP* (W-TCP), and TLS Protocols.

### W-HTTP

The W-HTTP is based on HTTP specification [2]. The goal of W-HTTP is to transfer the self-describing hypermedia resources.

The WAP terminal is turned to the HTTP server and the WAP proxy is turned to the HTTP client in order to achieve the push functionality. The W-HTTP supports the message body compression of responses in order to minimize the volume of data sent over the air. The W-HTTP supports the establishment of a tunnel using the CONNECT method in order to ensure end-to-end security.

### W-TCP

Some optimizations were provided to the TCP layer in order to support W-TCP implementations in a WAP configuration with a WAP HTTP proxy. As was shown in Section 6.4.2, the W-TCP must be supported in two modes, as follows:

- *Split TCP approach.* The TCP connection between the WAP client and the WAP proxy uses the W-TCP, there is a normal TCP connection between the WAP proxy and the origin server.

- *End-to-end TCP connection.* The W-TCP is used between the WAP client and the origin server without WAP proxy.

As TCP is not optimized for cellular network owing to high BER environments, long delays, and variable bandwidths, some enhancements were added in W-TCP, such as selective acknowledgment, path MTU discovery, large window size based on *bandwidth delay product* (BDP), among others (see [1]).

### Transport Layer Security

The TLS was introduced in order to provide end-to-end secure sessions between the WAP client and the HTTP servers. TLS tunneling is defined to support end-to-end security at transport level between a WAP client and an HTTP server over the W-TCP layer while using HTTP proxy. The WAP client creates a TLS tunnel by using an HTTP CONNECT method. A WAP profile that includes cipher suites, session resume, server authentication, client authentication, and certificate formats specifies the use of TLS to improve over the air efficiency.

### 6.4.5   Push Architecture

The push architecture is based on the client/server model. The push operation is initiated by the PI. This latter transmits a push content with delivery instructions to a *push proxy gateway* (PPG) using the *Push Access Protocol* (PAP). This one delivers the push content to the WAP client located in a handheld terminal using the Push *Over The Air* (OTA) Protocol. Figure 6.8 illustrates the push framework.

The PPG provides push connectivity between wired and unwired networks. The PAP is tunneled in HTTP1.1 between the PI and the PPG. The Push OTA Protocol runs on top of HTTP 1.1 (OTA-HTTP) or WSP (OTA-WSP) between the PPG and the WAP client.

The PPG supports several processing functions:

- *Push submission processing.* When the PPG receives a push message from a PI, it checks the acceptance of this message. Then the PPG translates the client address provided by the PI into a format understood by the mobile network. Next it may transform the push message content in order to prepare the OTA transmission by using a compact binary encoding like WBXML for transmission over OTA-WSP or by using a content encoding for transmission over OTA-HTTP. Next the PPG selects the appropriate Push OTA Protocol—the confirmed or unconfirmed push—and delivers the message. Note that the PI may submit a single push message addressed to multiple recipients; in this case the PPG expands the push message into multiple addresses for delivery.

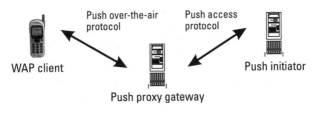

WAP client    Push over-the-air protocol    Push access protocol    Push initiator

Push proxy gateway

**Figure 6.8**   Push framework.

- *Result notification.* The PPG sends a final outcome of push submission (delivered, canceled, expired) if it was requested by the PI.

- *Status query.* The PPG gives the status of a previous push submission back to the PI on receipt of a request from the PI.

- *Client capability query.* The PPG sends back the client capabilities and preferences formatted as a UAProf document.

- *Delivery cancellation.* The PPG may support the cancellation of a pending push message if requested by the PI.

### 6.4.5.1   PAP

As the PAP was designed to be independent of the underlying transport layer, it may be used over any transport layer even if it was initially designed to run on top of HTTP 1.1.

The PAP supports the following operations:

- *Push submission.* The PI initiates a push submission toward the PPG.

- *Result notification.* The PI receives a final outcome of the push submission if it was requested.

- *Push cancellation.* The PI requests the PPG to cancel a previously submitted message.

- *Push replacement.* The PI requests the PPG to replace a previously submitted message with a new message.

- *Status query.* The PI requests the PPG to give back the status of a previously submitted message.

- *Client capabilities query.* The PI requests the PPG to give back the capabilities of a particular device on the network.

### 6.4.5.2   Push OTA Protocol

The Push OTA Protocol provides both connectionless and connection-oriented services for push contents between the PPG and the WAP client. The connectionless service (mandatory feature) always runs on the top of WSP. The connection-oriented service (optional feature) runs either on the top of WSP (OTA-WSP) or on the top of HTTP (OTA-HTTP). A push session is established for connection-oriented service using OTA-WSP; a TCP connection is established for connection-oriented service using OTA-HTTP.

*Session Initiation Request*

It is often necessary for the client to create a push session (OTA-WSP) or an active TCP connection (OTA-HTTP) when the PPG receives an asynchronous push submission in a connection-oriented service. An application called *session initiation application* (SIA) running on the WAP client side allows a PPG to request a WAP client for the establishment of a push or an active TCP connection. In order to activate the appropriate bearer, the SIA must listen to the *session initiation request* (SIR) from the PPG. The SIR is usually sent to the WAP client in a connectionless push by using a single SMS.

*OTA-WSP*

OTA-WSP is a protocol layer running on top of WSP. It is suitable for the receipt of push messages in a WAP client that do not support TCP/IP. It supports both connectionless and connection-oriented push services.

A WAP client may have two registered WDP ports for connectionless push: a nonsecure port (mandatory) and a secure port (optional) on which WTLS is supported. In case of connectionless push, the OTA-WSP is in charge of delivering the push content to the WAP client. The appropriate application is identified by an application identifier contained in the push content delivered on one of the registered WDP port in the WAP client.

A WAP client may support either secure or nonsecure transport services for connection-oriented push services. In this case, the OTA-WSP is in charge of delivering the push content to the WAP client. The appropriate application in the WAP client is identified by an application identifier contained in the push content delivered on the port defined during the session opening.

In order to create a push session for connection oriented push over OTA-WSP, the SIA must be implemented on the WAP client side and on the PPG side.

*OTA-HTTP*

OTA-HTTP is a protocol layer on top of HTTP. This feature was introduced in the WAP 2.0 release. It only supports connection-oriented push services. This means that an active TCP somehow must be established. For the delivery of a push content using OTA-HTTP, the PPG acts as a HTTP client, and the WAP client acts as HTTP server. This delivery of push content is accomplished by sending the HTTP POST method from the PPG to the WAP client.

There are two TCP connection methods for the creation of an active TCP listed as follows:

- *Terminal-originated TCP connection establishment method (TO-TCP).* The WAP client establishes a TCP connection toward the PPG.

- *PPG-originated TCP connection establishment method (PO-TCP).* The PPG establishes an active TCP connection toward the WAP client. This method is possible if the PPG has a way to know the IP address of the WAP client.

If the IP address is not available in the PPG, this latter cannot initiate the PO-TCP method. In this case, the PPG sends an SIR toward the WAP client so that the latter can establish an active TCP connection toward the PPG by initiating an IP connectivity with the mobile network.

The OTA-HTTP provides a way for the client to convey its CPI to the PPG. This latter may query at any time the CPI of the WAP client if an active TCP connection is established between the PPG and the WAP client. This registration operation is accomplished by sending an HTTP OPTIONS request from the PPG to the WAP client.

The OTA-HTTP provides a way for a mutual terminal/PPG identification and authentication. The authentication schemes referred to as "basic" (mandatory) and as "digest" (optional) are supported for PPG authentication by the WAP client. A similar mechanism is used by PPG for the WAP client authentication.

## 6.4.6   WTA Architecture

Figure 6.9 gives an overview of the WTA architecture.

### 6.4.6.1   WTA User Agent

A WTA framework is based on a dedicated WTA user agent. Unlike the WML agent, the WTA user agent has access to telephony services through a WTAI interface. This way the WTA user agent may control network functions (e.g., setting up calls) or device-specific features (e.g., phonebook). Moreover, the WTA user agent must react to network events (e.g., receipt of calls) and forward them to the browser. Access to WTA services is possible from the WML/WML2 language via a URL or from WMLScript by calling WTAI libraries functions.

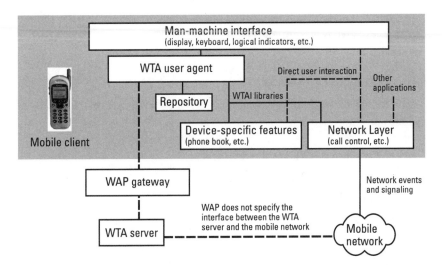

**Figure 6.9** WTA architecture. (*From:* [3]. © 2001, Wireless Application Protocol Forum, Ltd. All rights reserved.)

### 6.4.6.2 WTA Server

All WTA services are located on a WTA server. A WTA user agent references contents on a WTA server with associated URLs. A WTA server delivers content requested by a WTA user agent through a WAP gateway such as notification of incoming call, notification of new voice mails, or notification of call subscriber from message list or log.

***Note*** The WTA server is not specified by the WAP Forum.

### 6.4.6.3 Repository

A repository is a persistent storage module within the handheld terminal that contains several WTA contents. With this mechanism, the WTA user agent does not need to have access to the network for loading and executing of WTA service. Only the WTA user agent has access to the repository.

A repository contains channels linked to a set of resources. A resource is a network data object or service downloaded with WSP such as WML2 document, a WMLScript object, and a WBMP image. The channels are identified by a URL and have a freshness lifetime. A WTA event may be associated with a channel. This means that when a WTA event is detected, the user agent invokes the associated URL in order to activate the set of resources linked to this channel.

The WTA user agent is used to activate telephony functions in the terminal from WAP browser request and to interact with external events such as the receipt of a call.

### 6.4.6.4   Activation of WTA Service

There are several ways to initiate a WTA service via a WTA user agent, listed as follows:

- Access to a URL via the repository;
- Access to a URL via the WTA server;
- SI (push);
- Network event (transformed to WTA event in the client).

### 6.4.6.5   WTA Security

WTA security requirements have been defined in order to prevent any application from taking control of the mobile phone without user control. Each mobile network operator must provide an acceptable security level within its network. The user agents, except for the WTA user agent, may not have access to telephony services. As the WDP protocol provides several port numbers, it is required to use a specific secure port number for the WTA service, one not used by another WAE service. Another requirement is to submit the WTAI function calls performed by executables to user permission. This is achieved by defining several permission levels (blanket permission, context permission, single action permission) for each WTAI function.

### 6.4.6.6   Session Management

To facilitate interactions between a WAP gateway and a WTA server, WTA uses a WSP session, called WTA session. This latter may be established either in connection-oriented mode or in connectionless mode over a dedicated WDP port.

A WTA context makes it possible to associate several parameters with a given context (e.g., called numbers). One or several WTA sessions may be activated simultaneously within a WTA user agent. A session may be associated to a peculiar function. For example, one session is dedicated to call management and another one to voice mail.

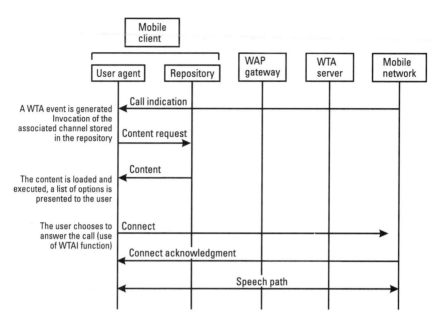

**Figure 6.10**  Incoming call selection.

6.4.6.7    Example of WTA Services

Figure 6.10 gives an example of incoming call detection. In this example, a valid channel associated with a WTA event is stored in the repository.

Figure 6.11 gives an example of a voice-mail application. In this example the user chooses to listen to one of the new voice mails.

## 6.4.7    Provisioning Architecture

There are two phases defined on the architecture for provisioning:

- Bootstrapping or bootstrap process;
- Continuous provisioning.

The bootstrapping is the initial configuration loaded in the WAP client, which contains connectivity information related to NAP, generic WAP proxies and a TPS. The connectivity information includes network bearers, protocols, and access point addresses as well as proxy addresses and TPS

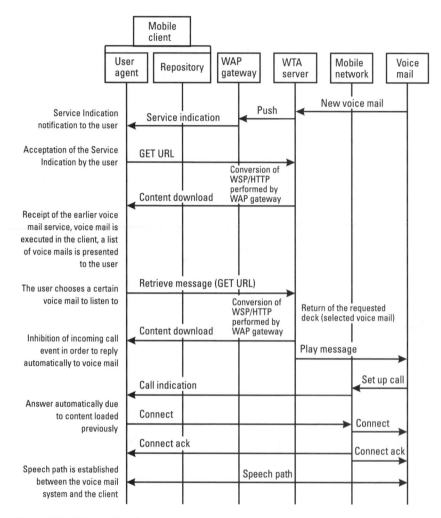

**Figure 6.11** Voice mail application.

URL or specific content locations URL. The goal of the bootstrapping is to create a trusted relationship between the WAP client and the WAP infrastructure with one or more trusted provisioning servers. This phase of provisioning is realized by the operator. It may be performed by various means such as a preconfigured SIM card, parameters manual entry; or OTA download.

The continuous provisioning allows occasional updating of configuration parameters in the WAP client. It is triggered either by a TPS or by a

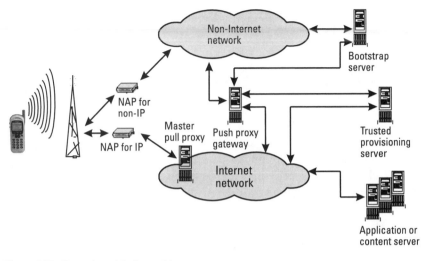

**Figure 6.12**   General provisioning architecture.

user. This provisioning phase is realized either by an operator or by a service provider. Before a continuous provisioning can be performed, a trusted relationship must be created between the WAP client and the WAP infrastructure. The context configuration updates are provided by a TPS which sends connectivity documents (XML document using a peculiar MIME type). The TPS identity (URL) is provided during the bootstrap process.

On the WAP client side, a provisioning user agent performs the provisioning methods and stores the connectivity information. Figure 6.12 shows the general provisioning architecture.

## 6.4.8   Security Architecture

### 6.4.8.1   Cryptography

*Secret Key Cryptography Mechanism*

The secret key cryptography mechanism provides a way to cipher a communication with a shared ciphering algorithm. A secret key is shared by the end-to-end users and is used for encryption and decryption. The secret key cryptography mechanism is also called a symmetrical cryptographic mechanism. Figure 6.13 shows the mechanism of secret key encryption.

There are two families of algorithms used with secret key encryption: stream cipher algorithms and block cipher algorithms. A block cipher

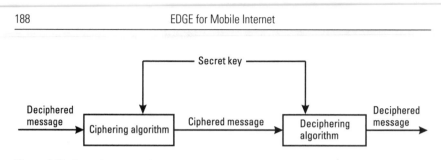

**Figure 6.13**  Secret key encryption.

enables enciphering an amount of data in one chunk. In order to prevent attacks, the level of security may be improved by increasing the key length. One mechanism to crack a key is to try all possible keys on a ciphered message. This is called exhaustive search. With this solution, the time to crack a key is $2^{\text{key size}} \times$ a software execution time per try. The WTLS layer provides various secret key encryptions as shown in Table 6.2.

**Note**  In order to avoid exhaustive search attacks, the WAP Forum advises to not use a key size of 40 bits.

*Public Key Cryptography Mechanism*

The public key cryptography mechanism provides each user with a pair of keys. The first key is secret, and known only by its owner. It is called the private key. The second key is public, and is called the public key. A message encrypted by a private key can be decrypted only by the associated public key, whereas a message encrypted by a public key can be decrypted only by the associated private key. This mechanism is based on the distribution of public keys. A public key may be computed easily from a private key while the reverse operation (i.e., the computation of a private key from a public key) is a very complex task requiring a time close to infinity.

**Table 6.2**
List of Secret Key Encryption Provided by the WTLS Layer

| Algorithm | Comments | Type of Cipher Algorithm | Key Size (bits) |
|-----------|----------|--------------------------|-----------------|
| RC5 | Rivest Cipher algorithm licensed by RSA | 64 bits block | 40 or 56 or 128 |
| DES | Data Encryption Standard designed by IBM | 64 bits block | 40 or 56 or 168 |
| IDEA | International Data Encryption Algorithm | 64 bits block | 40 or 56 or 128 |

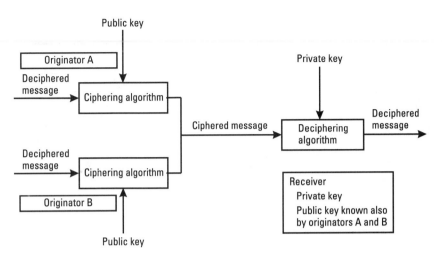

**Figure 6.14** Confidentiality with the use of a key pair (private key, public key).

The public key cryptography mechanism applies to an open network architecture since a public key is used by all originator entities for the encryption of messages toward an entity that uses the associated private key. Unlike the secret key cryptographic mechanism, the receiver entity does not need to send any peculiar key to each originator entity. Moreover, the encryption of a message by a private key enables the receiver entity to authenticate the originator entity with the associated public key. Figure 6.14 shows the use of a pair of keys (private key, public key) for confidentiality.

Figure 6.15 shows the use of a pair of keys (private key, public key) for authentication.

A combined use of private key and public key may be performed for the encryption of a message. An originator may encrypt a message first with its own private key, to be authenticated by the receiver, and then with the receiver public key to prevent reading by another receiver entity.

The algorithm for the public key mechanism is more complex than the one used for the secret key mechanism. Moreover, for the same key size a secret key mechanism is more secured than a public key mechanism. That is why a public key mechanism is combined with a secret key mechanism.

In case of a server client application, the server may send its public key to the client. The client generates a new secret key that is encrypted with the server public key and sent to the server. When the secret key is acknowledged by the server, the messages originated by the client are encrypted by using the

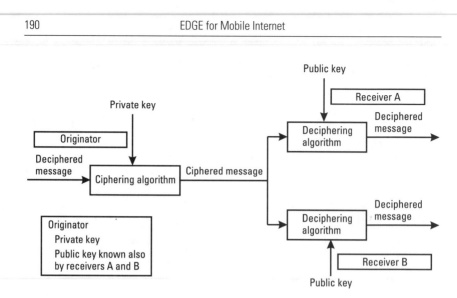

**Figure 6.15** Authentication with the use of a key pair (private key, public key).

secret key and are decrypted by the server by using the same secret key. Figure 6.16 shows a combination scenario for functional public and secret keys.

The WTLS layer combines public and secret keys for encryption. The server sends a public key to the client to be used either for the encryption of a random secret key or for the computation of a master secret key shared by both entities. The WTLS layer provides various public key encryption algo-

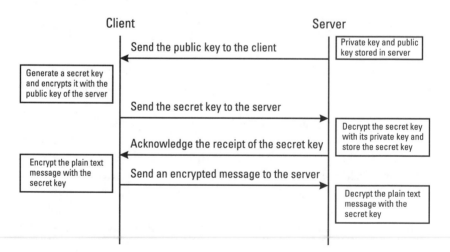

**Figure 6.16** Combination of public and secret keys.

rithms such as *Rivest Shamir Adleman* (RSA), *Diffie-Hellman* (DH), *Elliptic Curve Diffie-Hellman* (ECDH). The key size of these algorithms is variable. These public keys are used to yield a premaster secret key that enables computation of a master secret key owing to the use of two random values—one value generated by the server and the other by the client.

### 6.4.8.2 WIM Functionality

The WIM provides security functions for WTLS and for the application level security function. WIM is defined as a tamper-resistant module; it uses a physical hardware protection to fend off an attack on user data from an intruder.

The WIM functionality stores various data such as the supported encryption and authentication algorithms, the key pair for authentication and digital signature, its own certificate for each key pair, trusted certificates from a *certification authority* (CA), data related to WTLS sessions, and information on data protection with *personal identification numbers* (PINs). WIM performs signing operations, and cryptographic operations during the authentication step, and it analyses the received server certificates. Moreover, WIM generates the pre-master secret key as well as the master secret key. It provides the key material from the master secret key for the MAC and for the encryption keys.

WIM can be implemented in a smart card. It has been specified as an independent smart card application, meaning that it may be implemented either in a WIM-only card or in a multi-application card containing other applications. Thus, a WIM may or may not be embedded in an SIM card.

### 6.4.8.3 PKI Architecture

The PKI ensures the trusted relationships needed for server and client authentication. It is in charge of delivering certificates to the server and the client. Certificates are delivered by a trusted body called a CA. The certificate requests are sent toward a PKI portal that forwards the request to the CA.

A certificate is used to secure the process of public key distribution. This mechanism makes it possible to identify the entity that delivers a public key. A certificate is a message based on an X.509 format defined by the CCITT that contains its version, its issuer (e.g., Verisign organization), its validity period, the owner of the key, and the public key description. The CA appends a digital signature to the certificate that is known by all entities that need to authenticate the received certificates. When an entity needs to obtain its certificate, it must provide proof of its identity.

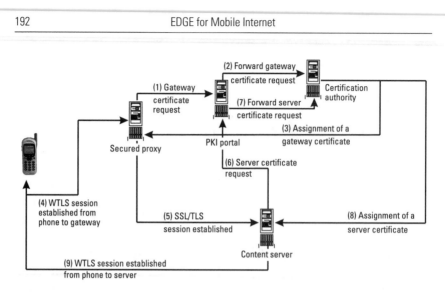

**Figure 6.17**  WAP PKI architecture for WTLS class 2.

The WAP PKI infrastructure reuses the principles of the PKI architecture and adapts them to the WAP environment.

In WTLS class 2, the gateway and the server provide a certificate to be authenticated by the client. The PKI infrastructure may provide a gateway certificate and a server certificate. The request for a short-lived certificate uses the HTTP GET method with an HTTP content type equal to "x-x509-user-cert." Figure 6.17 illustrates the delivery of a gateway and a server certificate.

In WTLS class 3, the client sends a certificate to the server in order to be authenticated when the client has to deliver its public key. A client certificate may be delivered by the PKI infrastructure. When the CA assigns a user certificate, it sends it either to the client or to a database to be stored. In the latter case, the CA also sends the user certificate URL to the client, which sends its public key with its user certificate URL to the server in order to be retrieved in the database. The architecture for the delivery of an user certificate is shown in Figure 6.18.

### 6.4.8.4   End-to-End Security Architecture

The WAP 2.0 specification has introduced an end-to-end security solution at the transport layer for end-to-end secure applications such as e-commerce, corporate access, and the like. The WAP 2.0 architecture with the use of HTTP and TLS stacks provides end-to-end secure sessions between the WAP client and the content server.

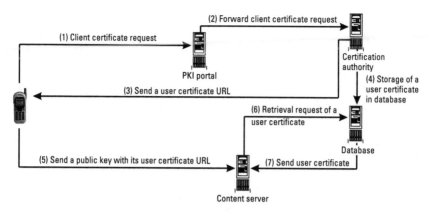

**Figure 6.18** WAP PKI architecture for WTLS class 3.

In the WAP 1.x architecture, there is a breaking off of the encrypting process in a gateway or in a proxy. In order to fill the end-to-end security requirement with a WAP 1.x architecture, a WTLS session is established between the client and a secure subordinate pull proxy. This latter is a pull proxy run within a secure domain—that is, a trusted network environment handled by a service provider. The WAP client may access to the secure subordinate pull proxy either directly, through a wireless port proxy, or via a secure NAP. Figure 6.19 shows the transport layer end-to-end security architecture.

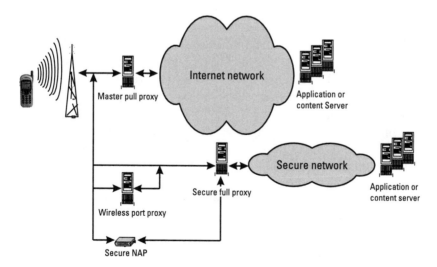

**Figure 6.19** Transport layer end-to-end security architecture.

The mechanism used to establish an end-to-end security session is as follows. First, the WAP client selects a URL and sends it to its default pull proxy by using the WSP get method. Then it is forwarded by the default pull proxy to the origin server. An HTTP status 300 response, as well as an XML navigation document are sent back by the origin server to the default pull proxy. The XML navigation document contains access parameters for a secure proxy for the selected URL. After the analysis of the navigation document, the default pull proxy forwards it with an HTTP error status to the WAP client. Then with this navigation path, the client establishes a WTLS session with the indicated secure proxy if no secure session already exists. The secure full proxy forwards the request to the origin server that replies to the WAP client via the secure full proxy.

### 6.4.9   Adapt Configuration End-to-End Architecture

The UAProf is transported over the Internet from the handheld terminal to the origin server through several proxies. The CC/PP exchange protocol is used to transport the CPI profile either over HTTP or over WSP. During the opening of an HTTP or WSP session, the WAP client transmits its CPI using a WBXML encoding according to the handheld terminal mode (WAP/WSP client or HTTP client). The WAP gateway or the HTTP proxy may add profile information to the CPI profile. The WAP gateway stores the CPI profile in its cache during the session lifetime in order to avoid requesting again the handheld terminal profile on the content server request. Then the WAP gateway or the HTTP proxy forwards the encoded CPI to the origin server. This latter parses the CPI profile, resolves the attribute values in order to generate an appropriate content. The CPI profile is stored persistently in a profile repository by the origin server. Finally, the origin server sends back its response to the handheld terminal by indicating whether the CPI has been taken into account.

In addition a PI has a way to adapt the content of its push message to the capabilities of a target device. The PI may request the PPG to retrieve the UAProf during the opening of an HTTP or WSP session by using a client query capability in the PAP. Then the PI may adapt the content of the subsequent messages to the handheld terminal by appending a capability entity that is formatted as a MIME multipart related message to the push message content. The PPG checks the profile requirement provided by the PI with the valid target device CPI. If there is a match, the push message is forwarded

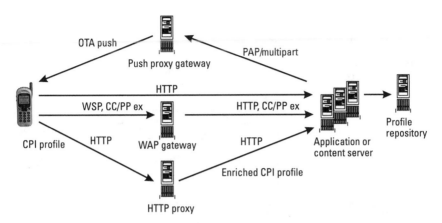

**Figure 6.20** Adapt configuration end-to-end architecture.

by the PPG over the Push OTA Protocol. The PPG may apply appropriate transformations to the push message content in order to achieve a match with the target device CPI. If there is no possible match, the PPG discards the push message. Figure 6.20 illustrates the adapt configuration end-to-end architecture.

## 6.4.10  MMS Architecture

### 6.4.10.1  MMS Framework

The architecture of MMS is composed of various elements, which are listed as follows:

- *MMS client.* The MMS user agent is implemented in the handheld terminal and is in charge of sending and receiving messages.
- *MMS proxy-relay.* This proxy interacts with the MMS client, has access to the MMS server for MM storage, and interacts with other messaging systems. It operates as an origin server (pull operations) or as a PI (push operations).
- *MMS server.* This server stores the MM messages.
- *E-mail server.* This server stores the Internet e-mail services.
- *Legacy wireless messaging system.* This system supports wireless messaging systems.

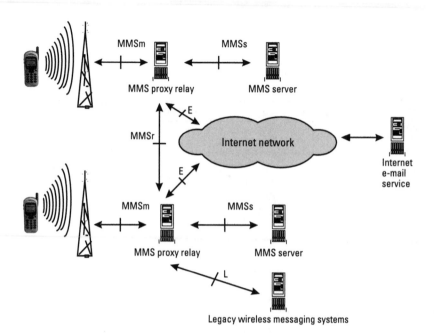

**Figure 6.21**  General MMS architecture.

Figure 6.21 gives the general MMS architecture.

The interface MMSm between the MMS client and the MMS proxy-relay is fully specified, unlike the other MMS architecture interfaces. A wireless transport such as WSP is used between the MMS client and the WAP gateway and HTTP is used from the WAP gateway to the MMS proxy-relay.

***Note***   The following interfaces are not yet standardized:

- MMSr interface between the MMS proxy-relay and other MMS systems;
- E interface between the MMS proxy-relay and Internet-based e-mail systems;
- L interface between the MMS proxy-relay and legacy wireless messaging systems.

### 6.4.10.2   MMS Addressing

The MMS specification has taken into account the constraints of limited keypad capacities in the wireless network in its addressing scheme. That is why the MMS specification has proposed to support a variety of addressing paradigms such as Internet e-mail address, MSISDN for GSM network, and receiver terminal IP address.

### 6.4.10.3  MMS Presentation

The MMS specifications include the concept of MMS presentation. This concept enables the sender of the *multimedia message* (MM) to describe multimedia animation including text, sounds, picture, and video to the receiving terminal. The WML language allows description of an MM presentation by offering sequencing and layout features. As for the *Synchronized Multimedia Integration Language* (SMIL), it provides extended capabilities such as multimedia object timing, and animation for MM presentations.

### 6.4.10.4  Security Aspects

MMS does not provide any specific mechanism to secure the transport of multimedia messages between an MMS client and the MMS proxy-relay. As MMS is an application layer, it may use, depending on its implementation, various mechanisms defined at the security layer level such as WTLS, WIM card, PKI, and secure S/MIME, which provides a way to encrypt the MIME components.

### 6.4.10.5  Content Adaptation

A content adaptation may be useful for reasons such as device capability constraints (e.g., limitations on content type), bandwidth considerations (e.g., data type inappropriate for a peculiar type of bearer), or roaming considerations (e.g., service constraints, pricing considerations) before delivering an MM to an MMS client. No specific content adaptation is specified within the MMS specifications. The WAP UAProf may be used to inform the MMS proxy-relay about the capabilities of the MMS client.

### 6.4.10.6  MMS Use Case

Figure 6.22 gives an example of MMS use case.

## 6.5  M-Services

The concept of M-Services allows the definition of a set of services with desired minimum requirements. The goal of this approach is to meet a successful mass-market launch of new services. This initiative was originated by

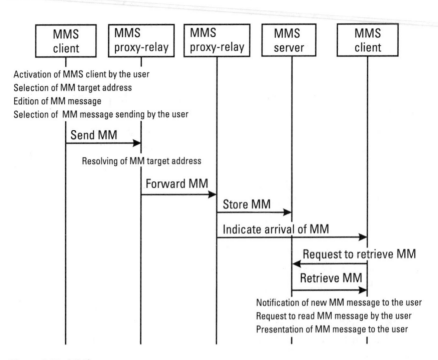

**Figure 6.22** MMS use case.

operators and was approved by the manufacturers, based on two observa-
tions: the difficult start of WAP services on available mobile browsers and the
tremendous success of iMode services in Japan brought about by NTT
DoCoMo. The success of the iMode concept stems from its structured
approach, which takes into account various aspects such as service aspects
and value chain. M-Services is not a standard, but it gives a feature guideline
for mobile phone design in order to achieve the same success as iMode.
M-Services is based on existing standards, more specifically on the WAP
1.2.1 standard. M-Services has defined two steps of feature conformance—
the first step for October 2001 and the second for June 2002.

M-Services gives guidelines on the following features:

- WAP 1.2.1 requirements (push, WTLS 2, WIM & WTLS3, UA
  Prof, WTAI Public);
- Handset features (GPRS/CSD, WAP provisioning);
- Download capability over WAP (WSP/HTTP GET, Download
  Fun);

- WAP graphical user interface (WML extension, dual-browser WML 1.X, and WML2);
- Download of media objects (ringing tone, picture, game, Java application);
- Messaging (SMS, EMS, e-mail, MMS);
- Additional services (SIM application toolkit, SyncML, bookmark).

# References

[1] Wireless Profiled TCP Specification.

[2] Fielding, R., et al., RFC2616 Hypertext Transfer Protocol—HTTP/1.1, June 1999.

[3] Wireless Telephony Application Specification.

# Selected Bibliography

Binary XML Content Format SIN 105.

Binary XML Content Format Specification.

End-to-End Transport Layer Security SIN 101.

End-to-End Transport Layer Security Specification.

External Functional Interfaces Specification.

General Formats Specification.

HTTP State Management SIN.

HTTP State Management Specification.

Persistent Storage Specification.

Provisioning Architecture Overview.

Provisioning Bootstrap SIN.

Provisioning Bootstrap Specification.

Provisioning Content Type SIN 003.

Provisioning Content Type SIN 004.

Provisioning Content Type Specification.

Provisioning User Agent Behavior.

Push Access Protocol SIN.

Push Access Protocol Specification.

Push Architectural Overview.

Push OTA Protocol SIN.

Push OTA Protocol Specification.

Push Message Specification.

Push Proxy Gateway Service SIN.

Push Proxy Gateway Service Specification.

Multimedia Messaging Service Architecture Overview.

Multimedia Messaging Service Client Transaction SIN 101.

Multimedia Messaging Service Client Transaction Specification.

Multimedia Messaging Service Encapsulation Specification.

Smart Card Provisioning Specification.

User Agent Profiling Specification.

WAP Client ID Specification.

WAP Cache Operation SIN.

WAP Cache Operation Specification.

WAP Caching Model Specification.

WAP Certificate Profile SIN 104.

WAP Certificate Profile Specification.

WAP Media Types Specification.

WAP Over GSM USSD Specification.

WAP Pictogram Specification.

WAP Pictogram SIN.

WAP Public Key Infrastructure SIN 103.

WAP Public Key Infrastructure Specification.

WAP Service Indication SIN.

WAP Service Indication Specification.

WAP Service Loading SIN.

WAP Service Loading Specification.

WAP Synchronization Specification.

WAP TLS Profile and Tunneling SIN 100.

WAP TLS Profile and Tunneling Specification.

WDP/WCMP Wireless Data Gateway Adaptation Specification.

Wireless Application Environment Specification.

Wireless Application Protocol Architecture Specification.

Wireless Control Message Protocol Specification.

Wireless Datagram Protocol Specification.

Wireless Identity Module Specification.

Wireless Markup Language Version 2 Specification.

Wireless Markup Language Version 1.3 Specification.

Wireless Markup Language Version 1.3 SIN.

Wireless Markup Language Version 1.3 SIN.

Wireless Profile Cascading Style Sheet Specification.

Wireless Profiled HTTP SIN 001.

Wireless Profiled HTTP Specification.

Wireless Session Protocol Specification.

Wireless Telephony Application Interface Specification.

Wireless Transaction Protocol Specification.

Wireless Transport Layer Security SIN 100.

Wireless Transport Layer Security Specification.

WML Transformations Specification

WMLScript Crypto API SIN 101.

WMLScript Crypto API Library Specification.

WMLScript Language Specification.

WMLScript Language SIN.

WMLScript Standard Libraries Specification.

WTAI, GSM Specific Addendum.

WTAI, IS-136 Specific Addendum.

WTAI, PDC Specific Addendum.

WTAI, IS95 Specific Addendum.

XHTML Mobile Profile Specification.

# List of Acronyms

| | |
|---|---|
| 3GPP | Third-Generation Partnership Project |
| 8-PSK | 8-state phase shift keying |
| AB | access burst |
| ABM | asynchronous balanced mode |
| ADM | asynchronous disconnected mode |
| AC | address control |
| ACK | acknowledgment |
| ADC | analog-to-digital converter |
| AGC | automatic gain control |
| AGCH | access grant channel |
| AM | amplitude modulation |
| AoCC | advice of charge—charging |
| AoCI | advice of charge—information |
| ARFC | absolute radio frequency channel |
| ARFCN | absolute radio frequency channel number |
| ARIB | Association of Radio Industries and Businesses |
| ARQ | automatic repeat request |
| AuC | authentication center |

| | |
|---|---|
| BC | bearer channel |
| BCCH | broadcast control channel |
| BCS | block check sequence |
| BDP | bandwidth delay product |
| BECN | backward explicit congestion notification |
| BER | bit error rate |
| BEP | bit error probability |
| BGIWP | barring of GPRS interworking profile |
| BH | block header |
| BLER | block error rate |
| BSN | block sequence number |
| BSC | base station controller |
| BSIC | base station identification code |
| BSS | base station system |
| BSSAP+ | base station system application part+ |
| BSSGP | Base Station System GPRS Protocol |
| BTS | base transceiver station |
| BVC | BSSGP virtual connection |
| BVCI | BSSGP virtual connection identifier |
| CA | certification authority |
| CAMEL | Customized Applications for Mobile Network Enhanced Logic |
| CAP | CAMEL application part |
| CC/PP | composite capabilities/preference profile |
| CCCH | common control channels |
| CDMA | code division multiple access |
| CDPD | cellular digital packet data |
| CGI | common gateway interface |
| CHAP | Challenge Handshake Authentication Protocol |
| C/I | carrier-to-interference ratio |

| | |
|---|---|
| CIR | channel impulse response |
| CKSN | ciphering key sequence number |
| CLNP | Connectionless Network Protocol |
| CLNS | connectionless network service |
| CO | cache operation |
| CONS | connection-oriented network service |
| CPFSK | continuous phase frequency shift keying |
| CPI | capability and preference information |
| CPM | continuous phase modulation |
| CPS | coding and puncturing scheme |
| CPU | central processing unit |
| CQC | client query capability |
| CS | coding scheme |
| CSD | circuit switched data |
| CSE | CAMEL service environment |
| CSI | CAMEL service information |
| CSS | cascading style sheet |
| CU | cell update |
| CUG | closed user group |
| CV | countdown value |
| CWTS | China Wireless Telecommunication Standard Group |
| D | direction |
| DAC | digital-to-analog converter |
| D-AMPS | Digital Advanced Mobile Phone Systems |
| dc | direct current |
| DCS 1800 | digital cellular system, GSM based on 1,800-MHz band |
| DES | data encryption standard |
| DISC | disconnect |
| DL | downlink |
| DLCI | data link connection identifier |

| DM | disconnected mode |
| DNS | domain name server |
| DP | detection point |
| DRX | discontinuous reception |
| DSC | downlink signaling counter |
| DTE | data terminal equipment |
| DHCP | Dynamic Host Configuration Protocol |
| E | extension |
| ECSD | Enhanced Circuit Switched Data |
| EDGE | Enhanced Data Rates for GSM Evolution |
| EDP-N | event detection point-notification |
| EDP-R | event detection point-request |
| EGPRS | Enhanced General Packet Radio Service |
| EIR | equipment identity register |
| EFI | external functionality interface |
| EFI AI | external functionality application interface |
| EMS | enhanced message service |
| ES/P | EGPRS supplementary/polling |
| ETSI | European Telecommunication Standards Institute |
| EVM | error vector magnitude |
| FACCH | fast associated control channel |
| FB | frequency correction burst |
| FBI | final block indicator |
| FCCH | frequency correction channel |
| FCS | frame check sequence |
| FDD | frequency division duplex |
| FDMA | frequency division multiple access |
| FECN | forward explicit congestion notification |
| FH | frame header |
| FH | frequency hopping |

| | |
|---|---|
| FN | frame number |
| FR | frame relay |
| FRMR | frame reject |
| FS | final segment |
| GGSN | gateway GPRS support node |
| GHOST | GSM hosted SMS teleservice |
| GMM | GPRS mobility management |
| GMSC | gateway mobile-service switching center |
| GMSK | Gaussian Minimum Shift Keying |
| GPRS | General Packet Radio Service |
| GPRS-CSI | GPRS-CAMEL subscription information |
| gprsSSF | GPRS service switching function |
| GSM | Global System for Mobile communications |
| GSMS | GPRS short message service |
| gsmSCF | GSM service control function |
| GSN | GPRS support node |
| GT | global title |
| GTP | GPRS Tunneling Protocol |
| GUTS | General UDP Transport Service |
| HCS | header check sequence |
| HDLC | High Level Data Link Control |
| HDML | Handheld Markup Language |
| HLR | home location register |
| HPLMN | Home Public Land Mobile Network |
| HSCSD | High Speed Circuit Switched Data |
| HSN | hopping sequence number |
| HTML | Hypertext Markup Language |
| HTTP | Hypertext Transfer Protocol |
| HTx | hilly terrain propagation channel, with speed $x$ km/hr |
| I | information |

| | |
|---|---|
| IDEA | International Data Encryption Algorithm |
| IDEN | Integrated Digital Enhanced Network |
| IF | intermediate frequency |
| IMAP | Internet Message Access Protocol |
| IMC | Internet Mail Consortium |
| IMEI | international mobile equipment identity |
| IMGI | international mobile group identity |
| IMSI | international mobile subscriber identity |
| IN | intelligent network |
| INAP | intelligent network application part |
| IP | Internet Protocol |
| IP3 | third order intercept point |
| IPCP | Internet Protocol Control Protocol |
| IPLMN | interrogating PLMN |
| IP-M | Internet Protocol multicast |
| IR | incremental redundancy |
| ISDN | Integrated Services Digital Network |
| ISI | intersymbol interference |
| ISL | input signal level |
| ISP | Internet service provider |
| LA | location area |
| LAC | location area code |
| LAI | location area identifier |
| LCP | Link Control Protocol |
| LI | length indicator |
| L2TP | Layer Two Tunneling Protocol |
| LLC | Logical Link Control |
| LMI | link management interface |
| LO | local oscillator |
| M | more |

| MA | mobile allocation |
| MAC | Medium Access Control |
| MAC | message authentication code |
| MAIO | mobile allocation index offset |
| MAP | Mobile Application Part |
| MCS | modulation and coding scheme |
| ME | mobile equipment |
| MIME | Multipurpose Internet Mail Extensions |
| MM | multimedia message |
| MMS | Multimedia Messaging Service |
| MS | mobile station |
| MSC | mobile-service switching center |
| MSK | minimum shift keying |
| MSISDN | mobile station ISDN |
| MT | mobile terminal |
| MTP | message transfer part |
| MTU | maximum transmission unit |
| NAP | network access point |
| NB | normal burst |
| NCP | Network Control Protocol |
| NER | nominal error rate |
| NF | noise factor |
| NM | network management |
| NMC | network management center |
| NNI | network-network interface |
| N-PDU | Network Protocol Data Unit |
| NS | network service |
| NSAPI | network service access point |
| NSC | network service control |
| NSDU | network service data unit |

| | |
|---|---|
| NSE | network service entity |
| NSEI | network service entity identifier |
| NSS | network subsystem |
| NV-VC | network service virtual connection |
| NV-VCI | network service virtual connection identifier |
| NS-VL | network service virtual link |
| NS-VLI | network service virtual link identifier |
| NZIF | near zero intermediate frequency |
| OOS | origin offset suppression |
| OSR | oversampling ratio |
| OSS | operator-specific service |
| OTA | over the air |
| PA | power amplifier |
| PACCH | packet associate control channel |
| PAGCH | packet access grant channel |
| PAP | PPP Authentication Protocol |
| PAP | Push Access Protocol |
| PBCCH | packet broadcast control channel |
| PC | power control |
| PCH | paging channel |
| PCCCH | packet common control channel |
| PCS1900 | Personal Communication System, GSM based on 1,900-MHz band |
| PCU | packet control unit |
| PDA | personal digital assistant |
| PDC | Personal Digital Cellular |
| PDCH | packet data channel |
| PDTCH | packet data traffic channel |
| PDN | packet data network |
| PDP | Packet Data Protocol |

| | |
|---|---|
| PDU | protocol data unit |
| PFI | packet flow identifier |
| PHS | Personal Handy Phone System |
| PI . | push initiator |
| PIM | personal information manager |
| PIN | personal identification number |
| PKI | public key infrastructure |
| PLL | phase locked loop |
| PLMN | Public Land Mobile Network |
| PNCH | packet notification channel |
| POP | Post Office Protocol |
| PO-TCP | PPG originated TCP connection establishment method |
| PPCH | packet paging channel |
| PPG | push proxy gateway |
| ppm | part per million |
| PPP | Point to Point Protocol |
| PR | power reduction |
| PRACH | packet random access channel |
| PSI | packet system information |
| PSK | Phase Shift Keying |
| PSTN | Public Switched Telephone Network |
| PT | payload type |
| PTCCH | packet timing advance control channel |
| P-TMSI | packet-temporary mobile station identity |
| PTP | point-to-point |
| PVC | permanent virtual circuit |
| QoS | quality of service |
| R | retry |
| RA | routing area |
| RAC | routing area code |

| RACH | random access channel |
|------|----------------------|
| RADIUS | remote authentication dial-in user service |
| RAI | routing area identity |
| RAS | remote access server |
| RAx | rural area propagation channel, with speed $x$ km/hr |
| RBSN | reduced block sequence number |
| RC | Rivest Cipher |
| RDF | Resource Description Framework |
| RF | radio frequency |
| RLC | radio link control |
| rms | root mean square |
| RNR | receiver not ready |
| RR | radio resource |
| RR | receiver ready |
| RRBP | relative reserved block period |
| RRM | radio resource management |
| RTI | radio transaction identifier |
| Rx | reception |
| RXLEC | received signal level |
| RXQUAL | received signal quality |
| SABM | set asynchronous balanced mode |
| SACCH | slow associated control channel |
| SACK | selective acknowledgment |
| SAP | service access point |
| SAPI | service access point identifier |
| SATK | SIM Application Toolkit |
| SB | synchronization burst |
| SCCP | signaling connection control part |
| SCH | synchronization channel |
| SCP | service control point |

| | |
|---|---|
| SDCCH | standalone dedicated control channel |
| SDU | service data unit |
| SGSN | serving GPRS support node |
| SI | service indication |
| SI | stall indicator |
| SI | system information |
| SIA | session initiation application |
| SIM | subscriber identity module |
| SIR | session initiation request |
| SL | service loading |
| SM | session management |
| SMIL | Synchronized Multimedia Integration Language |
| SMS | short message service |
| SMS-GMSC | short message service – gateway MSC |
| SMS-IWMSC | short message service – interworking MSC |
| SMTP | Simple Mail Transfer Protocol |
| SNDCP | Subnetwork Dependent Convergence Protocol |
| SN PDU | SNDCP PDU |
| SNR | signal to noise ratio |
| S/P | supplementary/polling |
| SS7 | Signaling System No. 7 |
| SSL | secured socket layer |
| SSP | service switching point |
| SVC | switched virtual circuit |
| TA | timing advance |
| TBF | temporary block flow |
| TCAP | transaction capabilities application part |
| TCH | traffic channel |
| TCP | Transmission Control Protocol |
| TDD | time division duplex |

| | |
|---|---|
| **TDMA** | time division multiple access |
| **TDP-R** | trigger detection point-request |
| **TFI** | temporary flow identity or identifier |
| **TI** | TLLI indicator |
| **TLLI** | temporary link level identity |
| **TLS** | transport layer security |
| **TMSI** | temporary mobile subscriber identity |
| **TO-TCP** | terminal originated TCP connection establishment method |
| **TOS** | type of service |
| **TPI** | transport information items |
| **TPS** | trusted provisioning server |
| **TRX** | transceiver |
| **TS** | training sequence |
| **TSC** | training sequence code |
| **TTA** | Telecommunications Technology Association |
| **TTC** | Telecommunication Technology Committee |
| **Tx** | transmission |
| **TUx** | typical urban propagation channel, with speed $x$ km/hr |
| **UAProf** | user agent profile |
| **UDP** | User Datagram Protocol |
| **UI** | unacknowledged information |
| **UL** | uplink |
| **UNI** | user-network interface |
| **URI** | Uniform Resource Identifier |
| **URL** | Uniform Resource Locator |
| **USF** | uplink state flag |
| **USSD** | unstructured supplementary service data |
| **UWCC** | Universal Wireless Communications Corporation |
| **VC** | virtual circuit |
| **VLR** | visitor location register |

| VPLMN | Visited Public Land Mobile Network |
| W3C | World Wide Web Consortium |
| WAE | wireless application environment |
| WAP | Wireless Application Protocol |
| WBMP | wireless bitmap |
| WBXML | Wireless Binary Extensible Markup Language |
| WCMP | Wireless Control Message Protocol |
| WDP | Wireless Datagram Protocol |
| W-HTTP | Wireless Profiled HTTP |
| W-TCP | Wireless Profiled TCP |
| WIM | wireless identity module |
| WML | Wireless Markup Language |
| WSP | Wireless Session Protocol |
| WTA | Wireless Telephony Application |
| WTAI | Wireless Telephony Application interface |
| WTLS | wireless transport layer security |
| WTP | Wireless Transaction Protocol |
| WWW | World Wide Web |
| XID | exchange identification |
| XHTML | Extensible Hypertext Markup Language |
| XML | Extensible Markup Language |
| ZIF | zero intermediate frequency |

# About the Authors

**Emmanuel Seurre** is a system engineer in Alcatel's handset division. He has worked on all of the GSM circuit data transmission technologies and has experience with all of the mobile layers related to GPRS and EDGE at the system and standards levels.

**Patrick Savelli** has acquired baseband and RF system expertise at the mobile phone divisions of Alcatel and Mitsubishi Electric, especially on GPRS, EDGE, and UMTS, for which he has followed the evolutions in the standards groups.

**Pierre-Jean Pietri** has worked for Alcatel, on a system specification team, and followed the standardization process for EDGE and GERAN for the BSS side. He now works for STMicroelectronics on the development of technical solutions for GSM/EDGE handsets.

# Index

*GSM and Personal Communications Handbook,* Siegmund M. Redl, Matthias K. Weber, and Malcolm W. Oliphant

*GSM Networks: Protocols, Terminology, and Implementation,* Gunnar Heine

*GSM System Engineering,* Asha Mehrotra

*Handbook of Land-Mobile Radio System Coverage,* Garry C. Hess

*Handbook of Mobile Radio Networks,* Sami Tabbane

*High-Speed Wireless ATM and LANs,* Benny Bing

*Interference Analysis and Reduction for Wireless Systems,* Peter Stavroulakis

*Introduction to 3G Mobile Communications,* Juha Korhonen

*Introduction to GPS: The Global Positioning System,* Ahmed El-Rabbany

*An Introduction to GSM,* Siegmund M. Redl, Matthias K. Weber, and Malcolm W. Oliphant

*Introduction to Mobile Communications Engineering,* José M. Hernando and F. Pérez-Fontán

*Introduction to Radio Propagation for Fixed and Mobile Communications,* John Doble

*Introduction to Wireless Local Loop, Second Edition: Broadband and Narrowband Systems,* William Webb

*IS-136 TDMA Technology, Economics, and Services,* Lawrence Harte, Adrian Smith, and Charles A. Jacobs

*Mobile Data Communications Systems,* Peter Wong and David Britland

*Mobile IP Technology for M-Business,* Mark Norris

*Mobile Satellite Communications,* Shingo Ohmori, Hiromitsu Wakana, and Seiichiro Kawase

*Mobile Telecommunications Standards: GSM, UMTS, TETRA, and ERMES,* Rudi Bekkers

*UMTS and Mobile Computing,* Alexander Joseph Huber and
 Josef Franz Huber

*Understanding Cellular Radio,* William Webb

*Understanding Digital PCS: The TDMA Standard,*
 Cameron Kelly Coursey

*Understanding GPS: Principles and Applications,*
 Elliott D. Kaplan, editor

*Understanding WAP: Wireless Applications, Devices, and Services,*
 Marcel van der Heijden and Marcus Taylor, editors

*Universal Wireless Personal Communications,* Ramjee Prasad

*WCDMA: Towards IP Mobility and Mobile Internet,* Tero Ojanperä
 and Ramjee Prasad, editors

*Wireless Communications in Developing Countries: Cellular and
 Satellite Systems,* Rachael E. Schwartz

*Wireless Intelligent Networking,* Gerry Christensen,
 Paul G. Florack, and Robert Duncan

*Wireless LAN Standards and Applications,* Asunción Santamaría
 and Francisco J. López-Hernández, editors

*Wireless Technician's Handbook,* Andrew Miceli

For further information on these and other Artech House titles,
including previously considered out-of-print books now available
through our In-Print-Forever® (IPF®) program, contact:

Artech House
685 Canton Street
Norwood, MA 02062
Phone: 781-769-9750
Fax: 781-769-6334
e-mail: artech@artechhouse.com

Artech House
46 Gillingham Street
London SW1V 1AH UK
Phone: +44 (0)20 7596-8750
Fax: +44 (0)20 7630-0166
e-mail: artech-uk@artechhouse.com

Find us on the World Wide Web at:
www.artechhouse.com